Applications and Metrology at Nanometer Scale 2

Reliability of Multiphysical Systems Set

coordinated by
Abdelkhalak El Hami

Volume 10

Applications and Metrology at Nanometer Scale 2

Measurement Systems, Quantum Engineering and RBDO Method

Pierre Richard Dahoo
Philippe Pougnet
Abdelkhalak El Hami

iSTE

WILEY

First published 2021 in Great Britain and the United States by ISTE Ltd and John Wiley & Sons, Inc.

ISTE Ltd
27-37 St George's Road
London SW19 4EU
UK

www.iste.co.uk

John Wiley & Sons, Inc.
111 River Street
Hoboken, NJ 07030
USA

www.wiley.com

Library of Congress Control Number: 2020950471

British Library Cataloguing-in-Publication Data
A CIP record for this book is available from the British Library
ISBN 978-1-78630-687-6

Contents

Preface

At the nanoscale, properties of matter cannot be explained by the laws of classical physics. To build models capable of interpreting the properties of matter on this scale, it is necessary to rely on the principles of quantum mechanics. The radical concepts of quantum mechanics and the development of nanotechnologies have contributed to the emergence of quantum engineering and a quantum information science.

Quantum engineering includes very sensitive materials and sensors that open up new fields of application, nanometer-sensitive measurement systems based on photonics and communication systems that perform well in terms of security. Quantum computing includes quantum computers and the development of new algorithms. Quantum computers are made up of quantum systems with two energy levels that follow the same laws of behavior as atoms or electrons enabling, with the development of quantum computing algorithms, performance that cannot be achieved with classical computers. Quantum technologies, nanotechnologies and nanoscience are identified as the sources of disruptive innovations that will bring technologies considered essential in the 21st Century.

The purpose of *Applications and Metrology at Nanometer Scale*, a book in two volumes, is to provide essential knowledge that will lead to the industrial applications of quantum engineering and nanotechnologies. The authors, through their skills and experience, combine their know-how in fundamental physics, engineering sciences and industrial activities. As with Volume 1, *Applications and Metrology at Nanometer Scale 2* is designed to provide applications for *Nanometer-scale Defect Detection Using Polarized Light* (Reliability of Multiphysical Systems Set Volume 2). It describes experimental and theoretical methods implemented in the framework of fundamental research to better understand physical–chemical processes at the nanoscale, presents examples of optical techniques based on the polarized light, allowing measurements to be made at the nanoscale, and illustrates the theoretical approaches with numerous applications.

This book is intended for master's and PhD students, engineering students, professors and researchers in materials science and experimental studies, as well as for industrialists of large groups and SMEs in the electronics, IT, mechatronics, or optical or electronic materials fields.

Chapter 1 deals with optical systems that enable measurements to be made on a nanoscale: the Fabry–Pérot cavity, homodyne interferometry, heterodyne interferometry, the optical lambda meter and ellipsometry with a rotating analyzer. The emphasis is on applications through exercises or analysis of study results on the use of interference techniques to study matter and materials.

Chapter 2 presents models of quantum physics that describe how a quantum two-energy level system interacts with its environment. As a free particle such as the electron that interacts with an external magnetic field with its spin, the derivation of the concept of spin from the Dirac equation is explained, which is the subject of an application exercise. The concept of density matrix (definition, propagation, equation of motion) is then presented and applied to a laser system with two energy levels and to a set of atoms interacting with the oscillating electric field of an electromagnetic wave. Finally, the Ising phenomenological model is presented, which is the subject of an application exercise.

Chapter 3 aims to provide theoretical foundations and examples of applications to understand the functioning of a quantum gate. A reminder is given on the modeling of light in quantum mechanics and on the representation by the Bloch sphere of the states of a two-level quantum system. The functioning of a quantum computer is introduced. Examples of applications show how to use the Bloch sphere, predict the evolution of an initial state of the system and obtain, by coupling, the oscillations of the Rabi population. Another application studies the coupling of an atom with light radiation and the effect on Rabi oscillations of a disagreement between the frequency of the atom and the frequency of the radiation. A final exercise deals with obtaining Ramsey fringes and their application to the functioning of a quantum gate.

Chapter 4 presents a reliability-based design optimization (RBDO) method of mechanical structures. This method guarantees a balance between the cost of defining the system and the assurance of its performance under the planned conditions of use. It is based on taking into account uncertainties and on the simultaneous resolution of two issues: optimizing the cost of producing structures performing the expected functions while ensuring a sufficient probability of operation under conditions of use (reliability). The RBDO method is applied to the optimization of the parameters of several mechanical components and of a printed circuit of an electronic board, and to ensure the reliability of the estimate of the measurement of the mechanical properties of carbon nanotube structures (Young's modulus of elasticity).

Pierre Richard DAHOO
Philippe POUGNET
Abdelkhalak EL HAMI
November 2020

Introduction

The scientific study of measurement is known as metrology. Any measure is based on a universally accepted standard and any measuring process is prone to uncertainty. In engineering science, measurement concerns various types of parameters. Legal metrology is imposed by a regulatory framework that the manufactured product must respect. Technical or scientific metrology involves the methods used to measure the technical characteristics of the manufactured product. In engineering sciences, measurement concerns various types of parameters. In a more general context of a systemic approach, metrology should also be considered in connection with other indicators of the production system. These measures enable the follow-up and development of the processes implemented for ensuring and optimizing product quality or reducing failure so that it meets client expectations. The ability of a product to meet quality and reliability expectations can be addressed in the design stage, according to a RBDO (Reliability-Based Design Optimization) approach described in Volume 2 of the Reliability of Multiphysical Systems Set, entitled *Nanometer-scale Defect Detection Using Polarized Light*. More generally, RBDO makes it possible to consider the uncertain parameters of manufacturing processes, measurement and operational conditions in order to optimize the manufacturing process, the design parameters and the overall quality of the product.

Nanometer-scale Defect Detection Using Polarized Light focused on three levels of design for manufacturing an industrial product:

– Numerical methods developed in engineering from mathematical models and theories in order to optimize product quality from its design according to RBDO. This methodology is a source of applications in engineering science intended to address optimization problems in the industrial field.

– Experimental methods developed in fundamental research relying on the light–matter interaction and on simulation-based analysis using theoretical models in order to make nanometer-scale measurements and conduct the analysis. These methods are used in nanosciences for the elaboration of knowledge leading to nanotechnologies.

– Finally, the application of these two approaches in the example presented in Chapter 9 of *Nanometer-scale Defect Detection Using Polarized Light* to the measurement of the physical properties of a nanomaterial, carbon nanotube.

In sciences, there are various ways to measure a dimension. The measuring instruments or methods employed depend on the scale at which metrology is approached. In order to describe the issues at stake for measurement at a given scale, we present the methods employed for the measurement processes at two scales of interest for scientists, namely the infinitely small, which corresponds to the Planck length of 1.6×10^{-35} m, and the infinitely large, which corresponds to the diameter of the Universe evaluated at 8.8×10^{26} m. This is to help the reader understand that, even though becoming an expert in a scientific field or in a given subject is not the objective, it is necessary to understand some basic tenets in order to master the methods used for successful metrology at a given scale.

In 1899, Planck determined a unit of length $l_P=\sqrt{(Gh/2\pi c^3)}\approx 1.6 \times 10^{-35}$ m, referred to as Planck length, based on fundamental constants: G, gravitational constant (6.6×10^{-11} Nm^2 Kg^{-2}), h, Planck's constant (6.64×10^{-34} Js) and c, the speed of light ($2.99{,}729{,}458 \times 10^8$ ms^{-1}). This length cannot be measured with the measurement technologies available on Earth. Indeed, the smallest length measurable at the LHC (Large Hadron Collider) of CERN, the particle accelerator in which two protons are made to frontally collide in a ring of 26,659 km, which led to the discovery in 2012 of the Higgs boson, is approximately 10^{-16} m, which is 19 orders of magnitude higher than the Planck length. CMS and ATLAS detectors were used in the observation of the Higgs boson, the latest prediction of the standard model not yet observed. The measurement at the scale of 10^{-16} m is made by compressing energy to reach an infinitely small spatial volume.

The principle of measurement at the scale of fundamental particles is mainly based on three relations: the de Broglie relation between the momentum p and the wavelength λ, $p=h/\lambda$, which introduces the wave–particle duality for matter; the relation that links the energy E of a particle to its wave frequency or wavelength λ, such as proposed by Einstein to explain the photoelectric effect $E = hc/\lambda$; and the relation that links the energy E of a particle of rest mass m to its rest mass energy and to its kinetic energy associated with its momentum $p=mv$, $E^2= m^2c^4 + p^2c^2$, as mentioned in Einstein's special theory of relativity. In the above formulas, v is the speed of the particle of mass m and c is the speed of light. The energy E can also be expressed by the formula $E= \gamma mc^2$, where γ is given by $\gamma =1/\sqrt{(1-v^2/c^2)}$. The speed of a particle is therefore given by $v/c=\sqrt{(1-(mc^2/E)^2}$.

In the LHC, the energy of a proton is 7 TeV ($1.2\ 10^{-6}$ J), far higher (by a factor of 7,500) than its rest energy, mc^2, which is 938 MeV. The formula for speed can then be rewritten as $v/c = (1-(m^2c^4/2E^2))$, which is equal to 1 to the nearest 10^{-8}. Using the relation $E= hc/\lambda$, the resulting value of the wavelength is of the order of 10^{-16} m, which gives the dimensions that can be reached in the LHC. The mass measured during two experiments at CERN in the LHC (8 TeV in 2012 and 13 TeV in 2015) is confirmed to the value of 125 GeV.

To detect the Higgs boson, a particle of mass 125 GeV associated with the Higgs field, while the mass of a proton is 938 MeV, the proton is accelerated and consequently its kinetic energy is increased so that its energy given by $E= \gamma mc^2$ significantly exceeds 938 MeV (8 TeV in 2012 and 13 TeV in 2015). The disintegration of colliding protons, each contributing an energy load of 8 TeV or 13 TeV, releases sufficient energy so that the Higgs boson can be expected to emerge during the recombination of subatomic particles. As the Higgs boson decays quasi-instantaneously after its emergence, the products of its decay must be analyzed to identify the excess energy and therefore the excess mass about 125 GeV.

It is worth noting that at the Planck length, the required energies that cannot be expected in a particle accelerator would lead to the emergence of black holes.

The opposite dimensional extreme towards the infinitely large corresponds to the spatial extent of the Universe, whose estimated value according to cosmologists is 10^{26} m. In cosmology, the observable Universe is a term used to describe the visible part of our Universe, the point from which light reaches us. It is a sphere whose limit is located at the cosmological horizon, having the Earth at its center. It is therefore a relative notion, as for other observers located somewhere else in the Universe, the observable sphere would not be the same (while its radius would be identical).

In cosmology, distances are measured in light-years. A light-year is the distance that light travels in one year, which corresponds to approximately 9.5×10^{12} m. The megaparsec, which is 3.26 million (3.26×10^6) light-years, is another unit of distance that is also specific to extragalactic astrophysics. Finding the size of the Universe involves accurate measurements of fossil radiation, or of the cosmic microwave background (CMB) radiation that originated in the Big Bang and can be used to determine the volume filled by the Universe since its creation. Predicted for the first time by Ralph Alpher in 1948 in his thesis work, CMB was discovered by Arno Penzias and Robert Wilson at "Bell Telephone Laboratories" during the development of a new radio receiver following the interferences detected independently of the orientation of the antenna they were building. While in a first approximation CMB is isotropic, accurate measurements of this radiation lead to determining H_0, the Hubble constant, which indicates the rate of expansion of the Universe.

In cosmology, detectors are above-ground telescopes. The WMAP (Wilkinson Microwave Anisotropy Probe) satellite launched in 2001 enabled the detection of CMB with good accuracy. Its intensity varies slightly in different directions of the sky and the fluctuations can be determined. Extremely accurate measurements of the WMAP in 2003 made it possible to calculate a value of H_0 of 70 kilometers per second and per megaparsec, which is within 5% in the hypothesis of a constant rate of expansion. Since the Universe is accelerating, during its expansion, the correction brought to H_0 made it possible to estimate the age of the Universe to 13.75 billion years, with a 0.1 billion margin of error. It is the scale fitting the domain to which corresponds the age of the Universe deduced from observations related to the Big Bang based on the inflationary model in an expanding Universe.

After the Big Bang, the elementary subatomic particles had no mass and could travel at the speed of light. After the expansion of the Universe and its cooling, the particles interacted with the Higgs field and consequently gained a mass.

In the history of the Universe, the elementary particles interacted with the Higgs field, 10^{-12} s after the Big Bang. The value of 125 GeV is considered as the critical value between a stable universe and a metastable universe. The "standard model of cosmology" elaborated at the beginning of this century, towards 2000, is probably at present the best model that enables the description of the evolution of the Universe, the significant stages in the history of the observable Universe as well as its current content, as revealed by astronomical observations. The standard model describes the Universe as an expanding homogeneous and isotropic space, on which large structures are overlaid as a result of the gravitational collapse of primordial inhomogeneities, which were formed during the inflation phase. There are still questions to be addressed, such as the nature of certain constituents of the Universe, black matter, and black energy and their relative abundance.

The inflationary model relies on the hypothesis of the Universe expanding with an exponential acceleration $R(t)=R_0\exp(H(t)t)$, 10^{-30} s after the Big Bang, where $H(t)$ is the Hubble constant. This constant is measured from the Doppler effect, which explains the red shift of the light radiation emitted by a distant star that is receding from the point of observation. The inflationary model allows for a plausible interpretation of the CMB isotropy, with relative variations of the measured temperature of 10^{-5}. Based on the data provided by the Hubble, COBE (Cosmic Background Explorer) and WMAP (Wilkinson Microwave Anisotropy Probe) telescopes, as well as by the BOOMerang (Balloon Observations Of Millimetric Extragalactic Radiation ANd Geophysics) and MAXIMA (Millimeter Anisotropy eXperiment IMaging Array) experiments, scientists were able to determine the age of the Universe is 13.75 billion light-years.

The Universe is presently in accelerated expansion: if its speed is 70 km/s at 1 Megaparsec, it doubles at 2 Megaparsec, reaching 140 km/s and so on. Considering the Doppler shift or the red shift for the receding stars, and therefore the fact that not only are the stars receding, but also those that are twice farther recede twice faster, therefore considering the metrics applicable to the space that is stretching while galaxies are receding, the 13.8 billion years between the beginning of the rapid expansion of the Universe 10^{-30} s after the Big Bang amount to 46.5 billion light-years, which is a radius of 93 billion light-years. Obviously, the light of stars that are at the periphery or at the cosmological horizon can no longer reach us, but as what we observe today goes back to the time needed for light to reach us while traveling a distance in a stretching space.

These two examples show that at each dimensional scale, besides the appropriate experimental measurement techniques required for observation, we must have a good mastery of the theories adapted for the interpretation and analysis of the gathered data. At each scale, the engineer must acquire specific knowledge elaborated in the laboratories and develop the competences to enable the mastery of technologies and the implementation of innovations.

This book which provides applications for Volume 2 of the Reliability of Multiphysical Systems Set (*Nanometer-scale Defect Detection Using Polarized Light*), focuses on knowledge elaborated at the nanometer scale for applications in the field of engineering sciences. The subjects approached are related to simulation experiments and engineering of nanometer-scale systems. The light–matter interaction has a special place among the subjects addressed, because the analysis of the properties and characteristics of matter is most often possible due to light being used as a probe. Similarly, simulation according to theoretical models based on quantum mechanics principles requiring field theory is also given particular attention.

Nanotechnologies and nanosciences are identified as sources of breakthrough innovations that will lead to the development of technologies that are considered primordial in the 21st Century. They should be deployed in eco-innovations and will increasingly become pervasive in the societal applications in various sectors. Without pretending to provide an exhaustive list, several examples are worth being mentioned: new energies and their recovery and storage, water purification, new materials that are lighter and more resilient for land and space transportation, construction and buildings, information technologies with quantum computers, embedded electronic systems and factory 4.0. The trend according to which states throughout the world offer financial support for the development of long-term projects in this field dates back to the beginning of the 21st Century. This is a reflection of the economic growth potential in nanotechnologies.

Similar to the inflationary model proposed by cosmologists to explain the countless galaxies and planetary systems, suns and black holes that constitute them, there was also a sharp increase in the volume of activities in nanosciences. The subjects approached in this book and in Volume 2 of the Reliability of Multiphysical Systems Set (*Nanometer-scale Defect Detection Using Polarized Light*) concern the field of engineers working in mechatronics, robotics and computation in modeling and simulation, for the societal spin-offs of nanotechnologies in the fields of land and space transportation, handicap, information and simulation technologies in a systemic approach. The level of knowledge acquired by the engineer should make innovation in nanotechnologies possible.

The contents of *Nanometer-scale Defect Detection Using Polarized Light* and *Applications and Metrology at Nanometer Scale 1 & 2*, jointly written by three authors, aim to develop knowledge that is essential at the nanometer scale, enabling trainee-engineers or engineers to develop nanotechnology-based devices or systems. To promote the deployment of nanotechnologies, the authors of these three books whose joint competences and experiences associate know-how in fundamental physics, engineering sciences and industrial activities cover a wide spectrum of application domains. *Nanometer-scale Defect Detection Using Polarized Light* builds a theoretical and experimental basis for understanding nanometer-scale metrology. This book in two volumes, *Applications and Metrology at Nanometer Scale*, enriches this theoretical basis with applications in the form of corrected exercises.

1

Measurement Systems Using Polarized Light

The Maxwell equations represent the physical phenomenon of light as an electromagnetic field which includes an electric field component and a magnetic field component oscillating in phase quadrature in a plane perpendicular to the direction of the light wave propagation. Photon wave–particle duality makes it possible to present both particle-like and wave-like behaviors. Interference is a phenomenon which is explained by the wave-like behavior of light. The wave-like properties of light and interference make it possible to study a number of physical phenomena, such as diffraction, interferometry, ellipsometry or holography. This approach has led to the development of innovative instruments using light–matter interaction and to various measuring systems. Mathematical approaches based on matrices are often used to calculate the effects of interference. These approaches use software such as MATLAB, MAPLE or computer programs written in Python, C language or Fortran. The finite element method can also be applied to design the instruments and simulate the expected observations.

1.1. Introduction

Chapters 3–6 of [DAH 16] present the theoretical characteristics of light. Light has dual properties; it is both a particle and a wave which is true also in the context of light–matter interaction. This chapter concentrates on applications through exercises or analysis of results of studies on the use of techniques using interference to study matter and materials.

The wave–particle duality in the case of a given particle (electron diffraction, neutron diffraction, etc.) is linked to the equation of L. de Broglie (1924) which relates the wavelength to the quantity of movement: $\lambda = h/p$, where p = mv, v being the speed and m the mass at rest. For a relativistic particle, with a speed close to that of light, the energy (E) of the particle is given by: $E^2 = p^2c^2+m^2c^4$. In the case of the photon which is a massless particle, $E = pc = hc/\lambda = h\nu$, because $c = \lambda\nu$, which leads to the equation used by Einstein (1905) to explain the photoelectric effect. When analyzed from quantum mechanics, the wave (p representation) or particle (E representation) behavior of the photon depends on the measurement or on the method used to detect light.

Optical instruments are essentially centered systems, the reference axis being the optical axis. To approach the homodyne and heterodyne interferometry techniques, the Handbury–Brown–Twiss experiment, the study of the Lyot Filter, Lidar, ellipsometry, spectroscopy based on polarized light and holography, the matrix method, presented in section 1.2 in the context of geometric optics and wave optics can be used. In wave optics, considering the two states of light polarization, Jones formalism makes it possible to model the effects of transmission plates using 2×2 matrices (Chapter 3 of [DAH 16]). The matrix formalism makes it possible to use digital tools, either by programming in advanced languages (Fortran, C, C ++, Visual Basic, Python, etc.) or by using, for example, software like MATLAB, SCILAB, MAPLE, COMSOL, ANSYS, NASTRAN and many others.

1.2. Matrix optics

In Chapter 3 of [DAH 16], the conditions for applying geometric optics to model the propagation of light are specified with respect to a characteristic length which is the wavelength related to the color of light. Under these conditions, the Snell–Descartes laws can be used (Figure 1.1, $n_1 sin i_1 = n_2 sin i_2$ and the same planes of incidence, reflection and refraction) to obtain the light ray propagation when it crosses optical elements: the diopter, the prism, convex and concave lenses and mirrors.

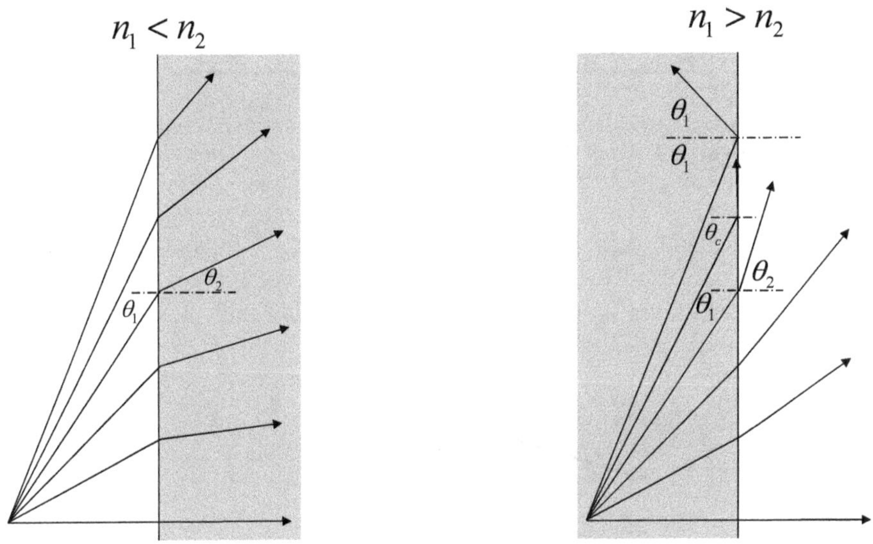

Figure 1.1. *Crossing of a plane diopter for $n_1<n_2$ and $n_1>n_2$. For a color version of this figure, see www.iste.co.uk/dahoo/metrology2.zip*

In Figure 1.1, the angles of incidence θ_1 and θ_2 verify the Snell–Descartes equation. When $n_1 > n_2$, there is the possibility of internal reflection when the angle of incidence is greater than the critical angle θ_c ($\sin \theta_c = n_2/n_1$).

A matrix approach can be applied to determine the trajectory of a light beam in the context of geometric optics. Usually, the optical systems used in instrumentation are axially symmetrical and only paraxial rays are considered which remain in the vicinity of the optical axis. Transfer matrices are used to locate the position of a ray as it propagates from one point to another. A ray of light can be defined by two coordinates: its position or height, r and its slope, $\theta = \partial r/\partial z$, relative to the optical axis, as shown in Figure 1.2.

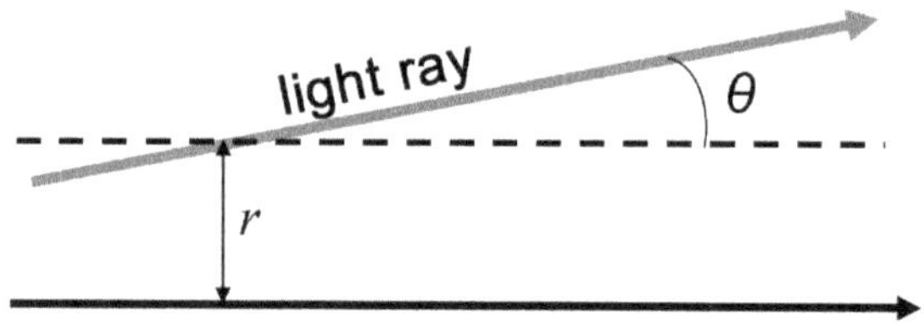

Figure 1.2. *Light radiation or vector radius of parameters r and θ. For a color version of this figure, see www.iste.co.uk/dahoo/metrology2.zip*

The transfer matrices, which represent the effect of translation, refraction or reflection, are used to locate the radius. The propagation of a ray in a homogeneous medium is shown diagrammatically in Figure 1.3.

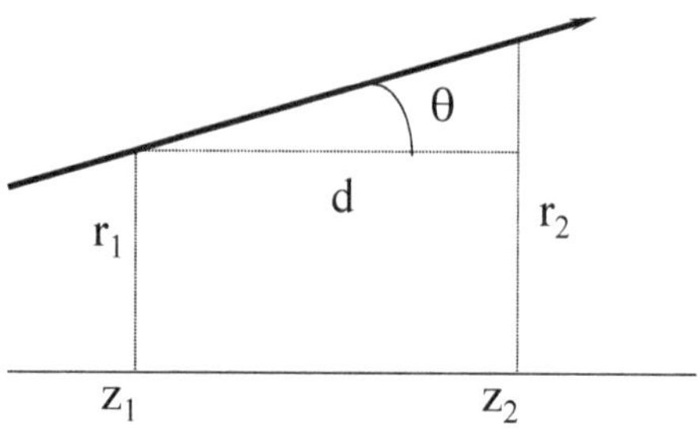

Figure 1.3. *Transfer matrix in an isotropic and homogeneous medium*

The light ray is identified by its distance r_i from the optical axis and the angle θ made by the tangent to the trajectory with the optical axis. For a homogeneous and isotropic medium, the propagation being rectilinear, the angle θ being a constant and having equation $r_2 = r_1 + \theta d$, the following matrix equation is obtained:

$$\begin{pmatrix} r_2 \\ \theta_2 \end{pmatrix} = \begin{pmatrix} 1 & d \\ 0 & 1 \end{pmatrix} \begin{pmatrix} r_1 \\ \theta_1 \end{pmatrix} \qquad [1.1]$$

where d is the distance between the two points parallel to the optical axis.

Similarly, the transfer matrix on a refractive surface from the characteristics, inclination of the rays, refractive indices of the media, distance from the optical axis is established as shown in Figure 1.4.

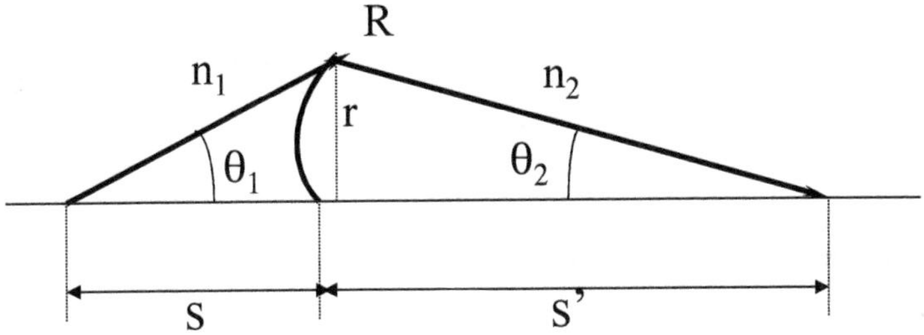

Figure 1.4. *Transfer matrix at the crossing of a diopter*

$r = r_1 = r_2$ is here a constant. In the Gaussian approximation of paraxial rays, the laws of geometric optics lead to: $-n_1/s + n_2/s' = (n_2-n_1)/R$, where R is the radius of curvature of the diopter. The relationship: $\theta_2 = \theta_1\ n_1/n_2 - (1- n_1/n_2)\ (r_1/R)$ is obtained, hence the matrix equation:

$$\begin{pmatrix} r_2 \\ \theta_2 \end{pmatrix} = \begin{pmatrix} 1 & 0 \\ \dfrac{n_2 - n_1}{n_2 R} & \dfrac{n_1}{n_2} \end{pmatrix} \begin{pmatrix} r_1 \\ \theta_1 \end{pmatrix} \qquad [1.2]$$

where $(1- n_1/n_2)\ (1/R)$ is the diopter focus.

Thus, to model the propagation of light in an optical system, a transfer matrix M is used, the elements of which denoted A, B, C, D characterize the optical system. This matrix describes the transmission of rays in optical components. For various common optical systems, the transfer matrices, which are given in Figure 1.5, are obtained.

Examples of optical assemblies are given in Figures 1.6 and 1.7.

The diameter of the beams is given in a ratio $D_2/D_1 = f_2/f_1$ in the telescope and in the case of Gaussian beams, the "waist" w is widened from the invariant: $w\theta = \lambda/\pi$.

Mirrors or cube corners can be used to reflect light and reflect back the beam (retro reflection) (Figure 1.8).

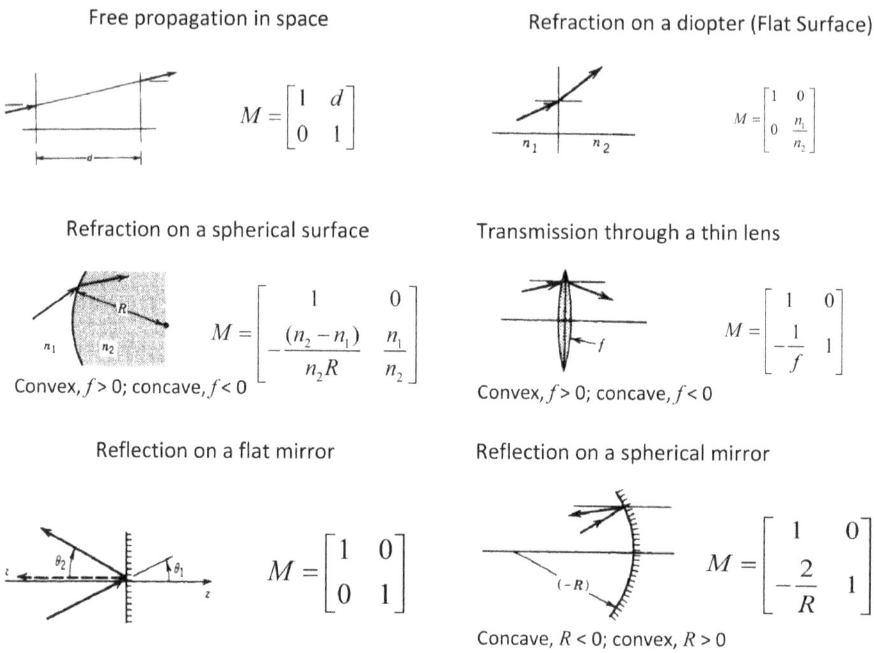

Figure 1.5. *Optical transfer matrices of different centered systems*

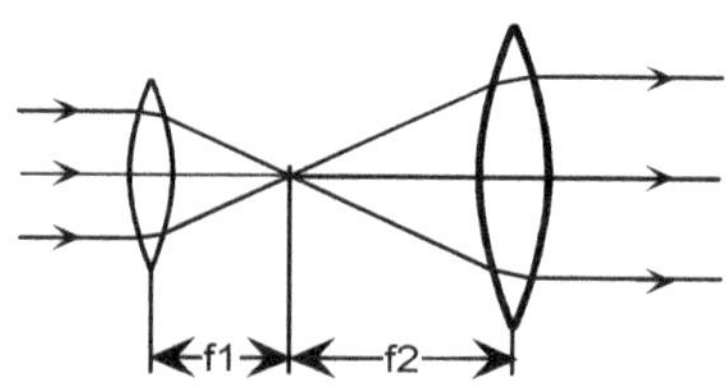

Figure 1.6. *Optical assembly for a telescope in a lidar*

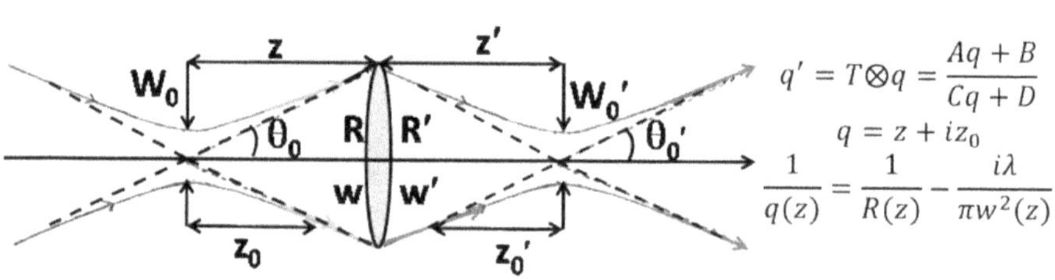

Figure 1.7. *Transmission of a Gaussian beam by a thin lens*

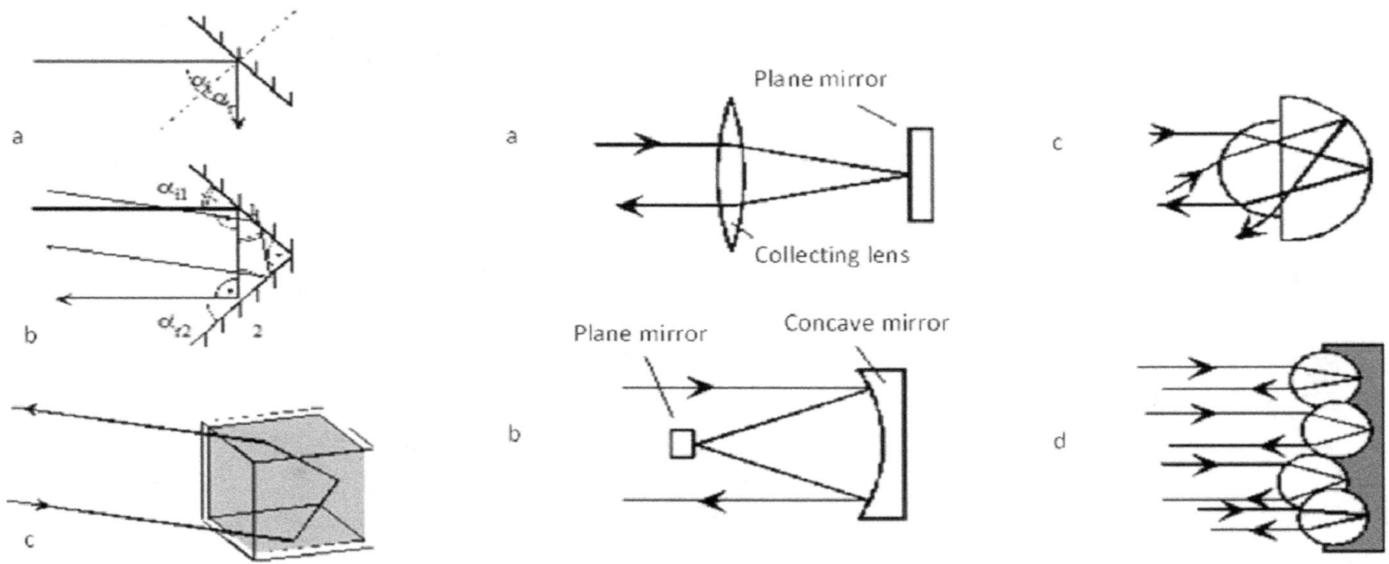

Figure 1.8. *Optical mounting of mirrors and retro-reflectors*

The propagation of light can be studied with 2×2 matrices in wave optics also by considering the two polarization states of light in Jones formalism. The optical axis is taken parallel to the Oz axis which corresponds to the direction of propagation of the wave and the plane of polarization of the wave is the Oxy plane, perpendicular to the direction of propagation since the electromagnetic wave is transverse. Any state of polarization can be described as a linear combination of two vibrations, as shown in Figure 1.9.

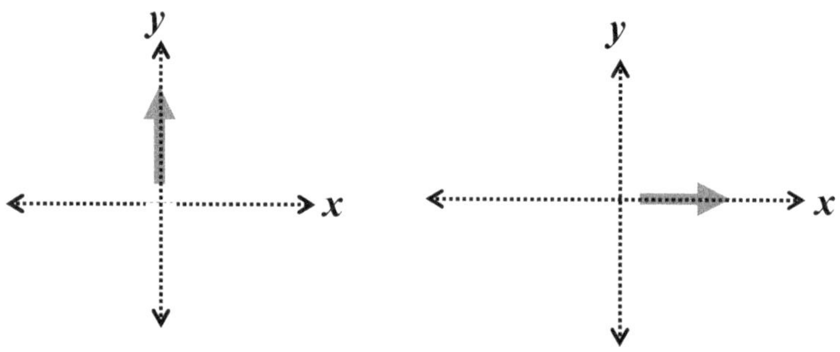

Figure 1.9. *Light polarization states. For a color version of this figure, see www.iste.co.uk/dahoo/metrology2.zip*

The mathematical expression is given by:

$$\vec{E} = E_{0x} e^{i(kz-\omega t+\varphi_x)} \hat{x} + E_{0y} e^{i(kz-\omega t+\varphi_y)} \hat{y}. \quad [1.3]$$

When the phases related to amplitudes and propagation are separated, the electric field of the wave is expressed as the product of a complex amplitude which contains all the information on polarization and a function which represents the propagation of the wave in the positive z direction and which gives the variation of the phase of the wave during its displacement on the trajectory of the light. The electric field of the polarized wave is written by:

$$\vec{E} = \left(E_{0x} e^{i\varphi_x} \hat{x} + E_{0y} e^{i\varphi_y} \hat{y} \right) e^{i(kz-\omega t)} \quad [1.4]$$

The composition of the electrical vibration of each of the perpendicular components as a function of the phase shift between the x and y components of the wave ($\Delta\varphi = \varphi_y - \varphi_x$) is shown for different situations in Figure 1.10.

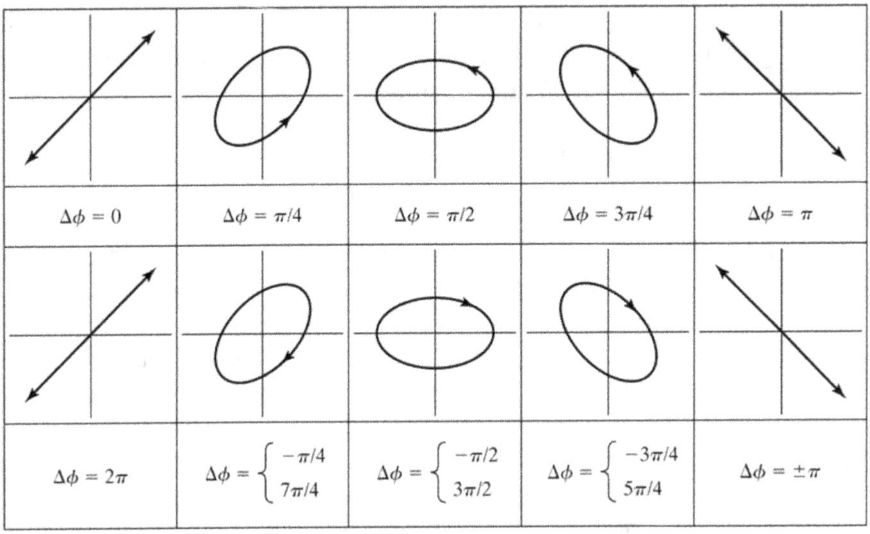

Figure 1.10. *Electric field vibrations in the polarization plane as a function of phase shifts*

In the case of a linear polarizer, all the vibrations of the E field are eliminated so that after the polarizer, there remains only one vibration which corresponds to a given direction which is the proper axis of the polarizer. Figure 1.8 shows an example of the effect of a linear polarizer.

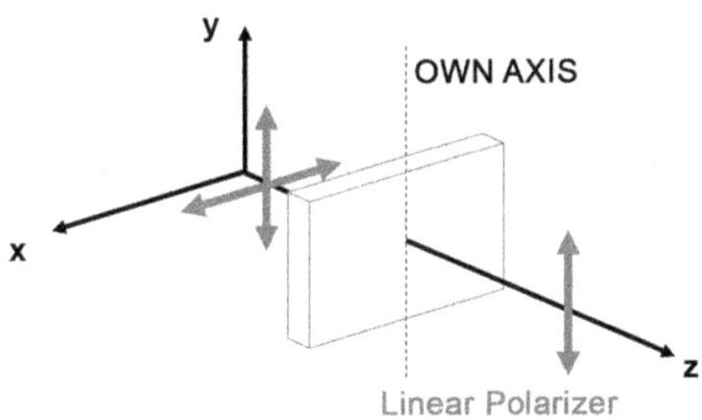

Figure 1.11. *Electric field vibration after a linear polarizer. For a color version of this figure, see www.iste.co.uk/dahoo/metrology2.zip*

In Jones formalism (Chapter 3 of [DAH 16]), the effect of the linear polarizer is described by a 2×2 transfer matrix, whose elements A, B, C and

D correspond to the modifications of the polarization state of the light. In the case of Figure 1.8, $A = B = C = 0$ and $D = 1$. The effect of the polarizer is to let the Ey component of the field pass according to:

$$\begin{pmatrix} 0 & 0 \\ 0 & 1 \end{pmatrix} \begin{pmatrix} E_x \\ E_y \end{pmatrix} = \begin{pmatrix} 0 \\ E_y \end{pmatrix} \quad [1.5]$$

The 2×2 matrices, which correspond to the effects of the various optical elements used in an experimental setup or a device, translate the actions of these optical elements. The wave is represented by the Jones vector with two components. Figure 1.11 gives the example of a phase retarder which introduces a phase difference between the two polarization components parallel to the own axes of the retarder (neutral axes of the birefringent material). In the figure on the right, it can be noticed that the higher the index, the more the wave is "packed" in the material. For $n = 2$, there are 3 and a half periods and for $n = 1.5$, there are 2 and a half periods over the thickness of the plate. The slow axis corresponds to the high index (more time is needed to cross the plate) or the lower phase speed and the fast axis to the lower index or the higher phase speed.

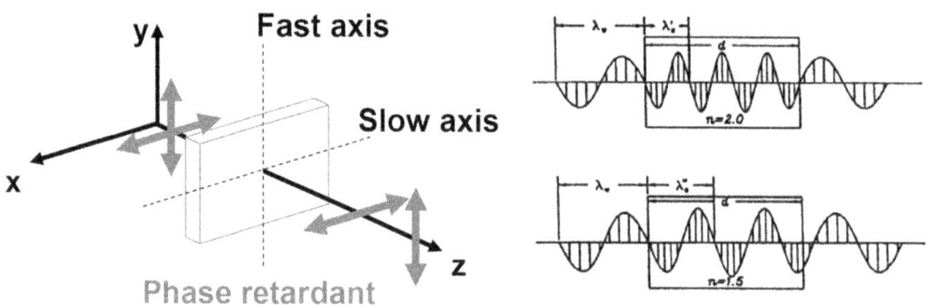

Figure 1.12. *Vibration of the electric field after a linear polarizer and difference in optical path along the axes. For a color version of this figure, see www.iste.co.uk/dahoo/metrology2.zip*

The effect of the self-timer is given by:

$$\begin{pmatrix} e^{i\varphi_x} & 0 \\ 0 & e^{i\varphi_y} \end{pmatrix} \begin{pmatrix} E_x \\ E_y \end{pmatrix} = \begin{pmatrix} E_x e^{i\varphi_x} \\ E_y e^{i\varphi_y} \end{pmatrix} \quad [1.6]$$

For example, for a thickness d, the phase shifts are: $\Delta\varphi_0 = -\frac{2\pi}{\lambda} n_0 d$ and $\Delta\varphi_e = -\frac{2\pi}{\lambda} n_e d$ on the ordinary and extraordinary axis, respectively. This leads to a phase shift between the paths of the wave along the ordinary axis and along the extraordinary axis of: $\Delta\varphi = \Delta\varphi_0 - \Delta\varphi_e = -\frac{2\pi}{\lambda}(n_e - n_0)d$.

If the phase difference $\Delta\varphi = \varphi_x\text{-}\varphi_y$ introduced between the two components x and y is $\pi/2$, the plate is said to be a quarter-wave plate. If this difference is π, the plate is said to be a half-wave plate. When the slow axis is vertical, the matrices are given by:

$$e^{-i\frac{\pi}{4}}\begin{pmatrix}1 & 0\\ 0 & i\end{pmatrix} \text{ and } e^{-i\frac{\pi}{2}}\begin{pmatrix}1 & 0\\ 0 & -1\end{pmatrix} \quad [1.7]$$

And when the slow axis is horizontal, by:

$$e^{i\frac{\pi}{4}}\begin{pmatrix}1 & 0\\ 0 & i\end{pmatrix} \text{ and } e^{i\frac{\pi}{2}}\begin{pmatrix}1 & 0\\ 0 & -1\end{pmatrix} \quad [1.8]$$

Figure 1.13 shows an example of a rotator that rotates the plane of polarization of the light at a certain angle. The rotator matrix is given by:

$$\begin{pmatrix}\cos\theta & -\sin\theta\\ \sin\theta & \cos\theta\end{pmatrix} \quad [1.9]$$

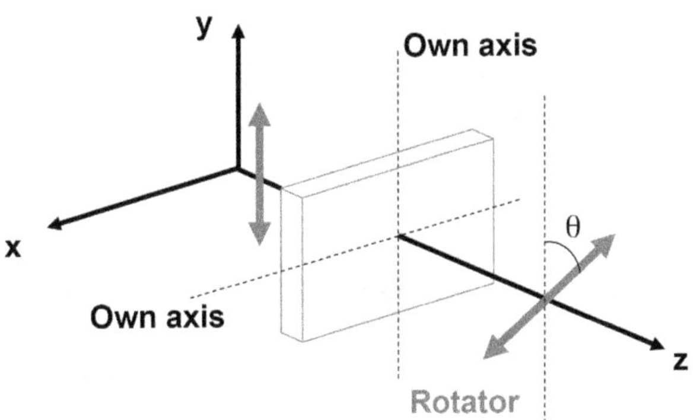

Figure 1.13. *Electric field vibration after a rotator. For a color version of this figure, see www.iste.co.uk/dahoo/metrology2.zip*

To model the effects of several devices on the state of polarization, it suffices to multiply the Jones vector of the input polarization by all the Jones matrices while respecting the order in which the elements act during propagation. The matrix of the first element is placed on the far right, first in the product, then the second element on the left of the first matrix and so on. Thus, a single Jones matrix, which is obtained by the product of the individual Jones matrices, describes the combination of several devices. Figure 1.14 gives a diagram of three devices that leads to $M = M_3M_2M_1$.

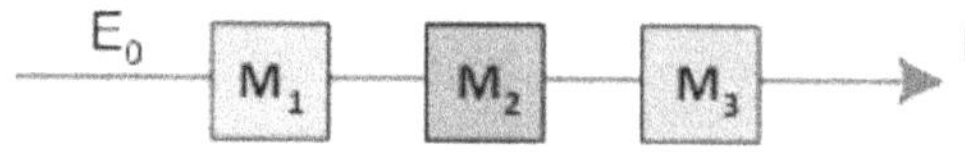

Figure 1.14. *Combination of three devices. For a color version of this figure, see www.iste.co.uk/dahoo/metrology2.zip*

In the case of crossed polarizers (Figure 1.15), we have:

$$\begin{pmatrix} E_{0x} \\ E_{0y} \end{pmatrix} \begin{pmatrix} 0 & 0 \\ 0 & 1 \end{pmatrix} \begin{pmatrix} 1 & 0 \\ 0 & 0 \end{pmatrix} = \begin{pmatrix} 0 & 0 \\ 0 & 0 \end{pmatrix} \begin{pmatrix} E_{0x} \\ E_{0y} \end{pmatrix} = \begin{pmatrix} E_{1x} \\ E_{1y} \end{pmatrix} = \begin{pmatrix} 0 \\ 0 \end{pmatrix} \qquad [1.10]$$

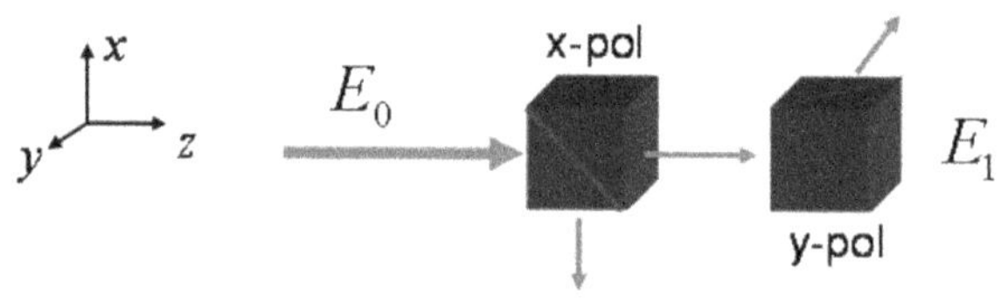

Figure 1.15. *Cross polarizers. For a color version of this figure, see www.iste.co.uk/dahoo/metrology2.zip*

1.3. Photon emission and detection

In interference devices, the light source is either a thermal source or a coherent laser source (Chapters 3 and 6 of [DAH 16]). Considering the states of creation and annihilation (section 3.4, Chapter 3 of [DAH 16]) and the quantum Hamiltonian expressed in terms of vibrators, the statistical operator in the canonical set is obtained in the form $\rho = \frac{\exp(-\beta H)}{Tr[\exp(-\beta H)]}$, where $\beta = 1/kT$ and H is the Hamiltonian operator whose expression is $H = \hbar\omega(a^+a + \frac{1}{2})$, i.e. $H = \left(\hat{n} + \frac{1}{2}\right)\hbar\omega$, where $\hat{n}$ is the operator number of photons. Considering a radiation field made up of a single mode of the photon, the probability of finding the system in the state with n photons $|n\rangle$ is given by the Bose Einstein statistic, i.e.: $P_n = \frac{\exp(-\beta n\hbar\omega)}{Z} = \exp(-\beta n\hbar\omega)$ $(1 - \exp(-\beta\hbar\omega))$, where Z is the partition function. The mean value of the distribution is given by $\langle n\rangle = \frac{1}{1-\exp(-\beta\hbar\omega)}$ and the standard deviation is given by: $\Delta n = \sqrt{\langle n\rangle + \langle n\rangle^2}$. The discrete probability distribution (Figure 1.16) can thus be expressed in the form: $P_n = \frac{(\langle n\rangle-1)^n}{(\langle n\rangle)^{n+1}}$.

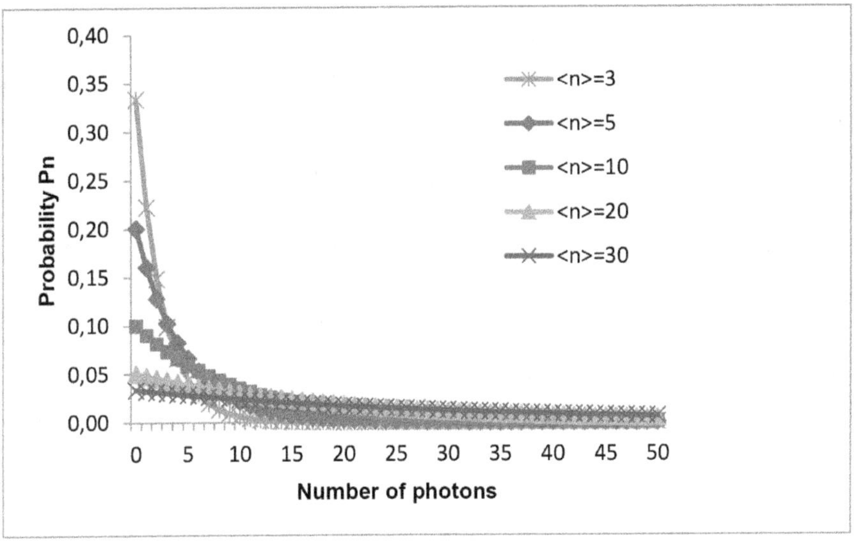

Figure 1.16. *Probability distribution for a thermal source. For a color version of this figure, see www.iste.co.uk/dahoo/metrology2.zip*

The Glauber state or the coherent state of an ideal single-mode laser is given by (equation [3.65], Chapter 3 of [DAH 16]) $|\alpha\rangle$, an eigenstate of the annihilation operator a of the photon, of eigenvalue α. As a is a non-hermitic operator, the phase α is a complex number which corresponds to the complex amplitude of the wave in classical optics of an electromagnetic field [GLA 63, SUD 63, GLA 67, ARE 72, DAV 96]. $|\alpha\rangle$ can be written in the basis of kets $|n\rangle$ of the Fock space as:

$$|\alpha\rangle = e^{-|\alpha|^2/2} \sum_{n=0}^{\infty} \frac{\alpha^n}{\sqrt{n!}} |n\rangle \equiv |re^{i\theta}\rangle.$$

This equation relates the wave nature, to the corpuscular nature of light, and shows that in the coherent state, the number of photons is not defined while the phase is (except for the uncertainty equation). These coherent states represent quasi-classical states insofar as a phase θ and an average amplitude *r* are associated with them.

The probability of having n photons in a coherent state $|\alpha\rangle$ is $P_n = |\langle n|\alpha\rangle|^2$. It can be shown that the distribution of photons follows a Poisson law (Figure 1.17): $P_n = e^{-|\alpha|^2}\frac{\alpha^{2n}}{n!}$, where the term $|\alpha|^2$corresponds to the mean value of the number of photons $\langle n\rangle$ associated with the operator $N(n) = a^\dagger a\rangle$. This mean value is given by $\langle n\rangle = \langle\alpha|N(n)|\alpha\rangle = |\alpha|^2$and the variance by:

$$\sigma^2 = \langle\alpha|N^2(n)|\alpha\rangle - |\langle\alpha|N(n)|\alpha\rangle|^2 = |\alpha|^2,$$

so that the standard deviation is given by $|\alpha|$. Thus:

$$P_n = e^{-\langle n\rangle}\frac{\langle n\rangle^n}{n!}.$$

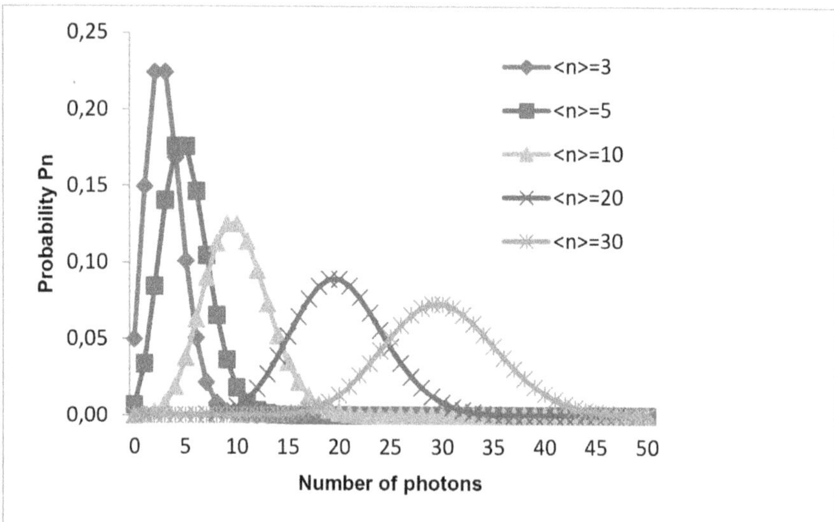

Figure 1.17. *Probability distribution for a coherent source. For a color version of this figure, see www.iste.co.uk/dahoo/metrology2.zip*

Photon detection is the result of a light–matter interaction. The electromagnetic field of a light wave has an electrical component E and a magnetic component B arranged perpendicular to each other while being perpendicular to the direction of propagation of the wave. The Poynting vector $\boldsymbol{S}$ of such a wave represents the power transported by the unit of surface and is expressed by: $\vec{S} = \frac{1}{2}(\vec{E} \wedge \vec{H}^*)$, where $|\vec{H}| = \frac{\vec{E}}{Z_0}$, where Z_0 is the characteristic impedance of the medium, which in the vacuum is equal to $\sqrt{(\mu_0/\varepsilon_0)}$. The Poynting vector is parallel to the direction of wave propagation in a homogeneous isotropic medium (Figure 1.17). Detection of the energy carried by a wave is done by placing the surface of a detector perpendicular to the path of the wave (Figure 1.18).

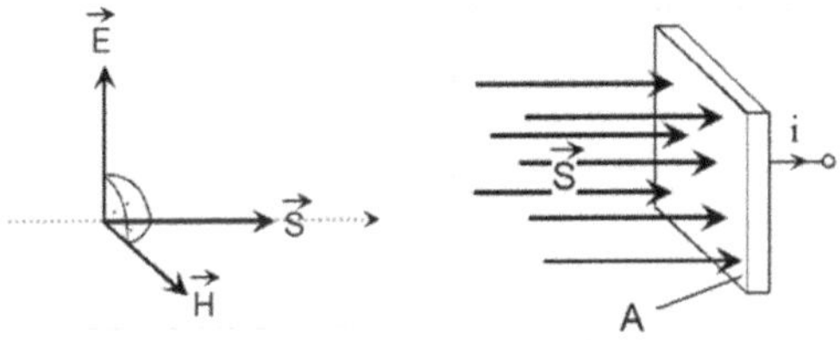

Figure 1.18. *Light detection*

The total power P carried by the wave and detected on a surface A is given by: $P = \oiint \vec{S}\vec{n}dxdy$ which leads to $P = \frac{1}{2Z_0}\oiint|\vec{E}|^2 dxdy = \frac{A}{2Z_0}|\vec{E}|^2$.

A detector being quadratic, if R is the response of the detector, the photo-current is given by: $i = RP = R\frac{A}{2Z_0}|\vec{E}|^2 = K|\vec{E}|^2$. As the intensity I of the light is given by $I = |\vec{E}|^2$, the following relation is obtained: for the photo-current: $i = K|\vec{E}|^2 = KI$. Detection is a nonlinear process.

Let W be the energy incident on the detector. The number n of incident photons on the detector is given by: $n = \frac{W}{W_{ph}} = \frac{W}{h\nu}$. The total charge is equal to: $q = \eta e\frac{W}{h\nu}$, where η is the detection efficiency. The photo-current is given by:

$$i = \frac{dq}{dt} = \frac{\eta e}{h\nu}\frac{dW}{dt} = \frac{\eta e}{h\nu}P = RP.$$

1.4. Application exercises on interferometry

In its operating principle, an interferometer is a two-wave interference optical device, which is based on the separation of a wave (Figure 1.19) by amplitude or wave-front splitting. The Michelson interferometer is used for interference experiments. It was made famous following the experiments of Michelson–Morley [MIC 87] on the verification of the existence of the ether in an approach based on high resolution measurement of the phase to determine a possible Doppler effect on the speed of the light.

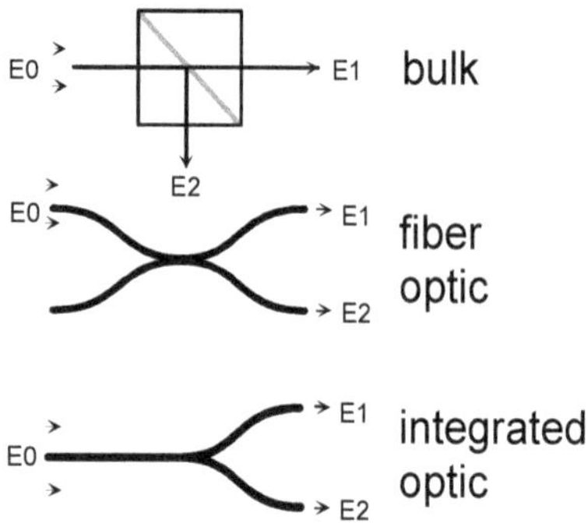

Figure 1.19. *Beam splitter devices*

In operation, an incident ray from a light source is partially reflected on the separator towards the mirror M_1 and partially transmitted towards M_2. After reflection on the two mirrors, the rays on paths 1 and 2 pass again through the separator. The rays coming out of the interferometer after having traveled different optical paths interfere and, since they are coherent, coming from the same wave surface, give rise to light interference.

The device consists of a fixed separator, and an adjustable compensator using adjustment screws, to orient it parallel to the separator. The light beam, which separates into two paths, is reflected on two mirrors, M_1 and M_2. The mirror M_2, facing the source in Figure 1.20, is mobile and can be translated and oriented using a three-screw system (line, point, plane system) by a so-called coarse adjustment. The parallelism between the mirrors M_1 and M_2 is obtained in a fine manner by fine adjustment screws on the fixed mirror.

Depending on the path difference (Δx) between the two rays which depends on the relative position of the two mirrors M_1 and M_2, the interference is constructive or destructive. The fact that the rays coming from M_2 cross twice the separator whereas those coming from M_1 cross it once induces additional phase shifts which are compensated by a compensator, which symmetrizes the optical paths in the two arms of the Michelson.

After reflection on each of the mirrors, the beam, which is recombined again at the level of the intensity separator cube, is sent to a detector which delivers an electrical signal to a counter whose role is to determine the number of fringes which pass as and measurement of the displacement of the movable mirror. A fringe is a whole cycle of varying light intensity, going from light to dark and back to light. Each cycle corresponds to a displacement of $\lambda_0 = 2n$. Knowing λ_0 and n, one deduces displacement from it.

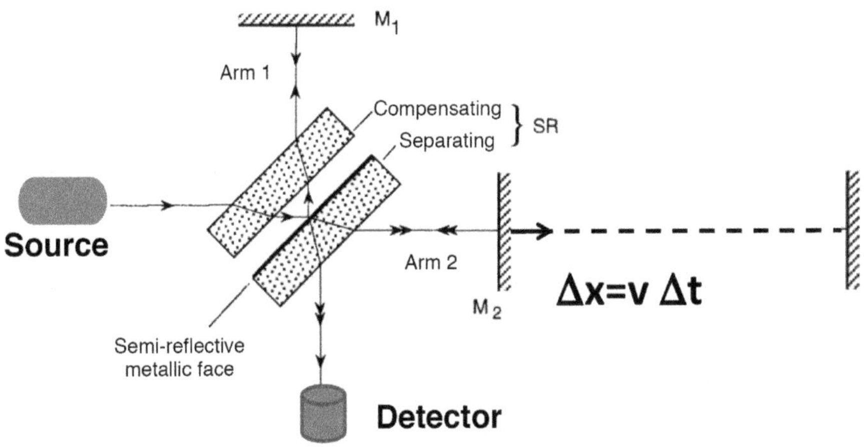

Figure 1.20. *Diagram of an interferometer. For a color version of this figure, see www.iste.co.uk/dahoo/metrology2.zip*

Interferometers are used in spectroscopy to identify molecules that absorb light in the IR domain (Reliability of Multiphysical Systems Set, Volume 1 [CAR 16]).

1.4.1. *Propagation of electromagnetic waves in a Fabry–Pérot cavity*

Two concave mirrors with radii R_1 and R_2 are considered (Figure 1.21). For paraxial rays, for which all the angles are small, the relation between (y_{m+1}, θ_{m+1}) and (y_m, θ_m) is linear and can be written in a matrix form by:

$$\begin{bmatrix} y_{m+1} \\ \theta_{m+1} \end{bmatrix} = \begin{bmatrix} A\,B \\ C\,D \end{bmatrix} \begin{bmatrix} y_m \\ \theta_m \end{bmatrix}$$

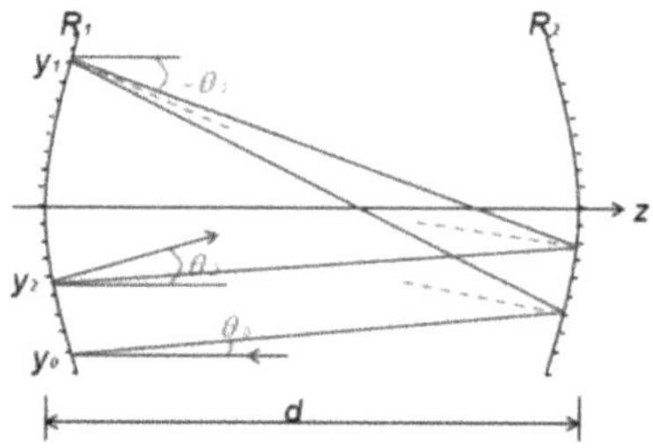

Figure 1.21. *Diagram of a Fabry–Pérot cavity. For a color version of this figure, see www.iste.co.uk/dahoo/metrology2.zip*

A ray which starts from the point m, (y_m, θ_m) of the mirror M_1 (radius of curvature R_1), propagates in the vacuum over a distance d to then arrive at the mirror 2 where it undergoes a reflection on M_2 is considered. It then starts again from M_2, propagates in the vacuum over a distance d and is reflected on M_1 at the point (y_{m+1}, θ_{m+1}).

1.4.1.1. *Question*

Demonstrate from the parameters of the cavity that the elements of the matrix M are expressed by:

$$A = 1 + {}^{2d}\!/_{R_2}$$

$$B = 2d(1 + {}^{d}\!/_{R_2})$$

$$C = {}^{2}\!/_{R_1} + {}^{2}\!/_{R_2} + {}^{4d}\!/_{R_1 R_2}$$

$$D = 1 + {}^{4d}\!/_{R_1} + {}^{2d}\!/_{R_2} + {}^{4d^2}\!/_{R_1 R_2}$$

1.4.1.2. *Solution*

The matrix M, which makes it possible to pass from the mirror of radius R_1 to the mirror of radius R_2, is obtained by multiplying the matrices corresponding to the path from mirror 1 to mirror 2 (M_1), then the reflection on the mirror of radius R_2 (M_2), then the path from mirror 2 to mirror 1 (M_3) and, finally, the reflection on mirror 2 (M_4).

Thus:

$$\begin{bmatrix} y_{m+1} \\ \theta_{m+1} \end{bmatrix} = M4M3M2M1 = \begin{bmatrix} A\,B \\ C\,D \end{bmatrix} \begin{bmatrix} y_m \\ \theta_m \end{bmatrix}$$

with:

$$M1 = \begin{bmatrix} 1 & d \\ 0 & 1 \end{bmatrix}, M2 = \begin{bmatrix} 1 & 0 \\ \dfrac{2}{R_1} & 1 \end{bmatrix}, M3 = \begin{bmatrix} 1 & d \\ 0 & 1 \end{bmatrix} \text{and } M4 = \begin{bmatrix} 1 & 0 \\ \dfrac{2}{R_2} & 1 \end{bmatrix}.$$

We can therefore write that the matrix M has the expression:

$$M = M4M3M2M1 = \begin{bmatrix} 1 & 0 \\ \dfrac{2}{R_2} & 1 \end{bmatrix} \begin{bmatrix} 1 & d \\ 0 & 1 \end{bmatrix} \begin{bmatrix} 1 & 0 \\ \dfrac{2}{R_1} & 1 \end{bmatrix} \begin{bmatrix} 1 & d \\ 0 & 1 \end{bmatrix} = \begin{bmatrix} 1+\dfrac{2d}{R_2} & 2d(1+\dfrac{d}{R_2}) \\ \dfrac{2d}{R_1}+\dfrac{2d}{R_2}+\dfrac{4d}{R_1R_2} & 1+4d/R_1+2d/R_2+4d^2/R_1R_2 \end{bmatrix}$$

1.4.2. *Propagation of electromagnetic waves in a material*

A retarder plate of thickness d is considered (Figure 1.11). This plate is characterized by main axes, a fast axis (A_R) of index nr and a slow axis (A_L) of index n_l.

1) What do these axes represent?

Solution:

The slow and fast axes correspond to the own axes of the plate which are determined by the symmetry properties (see Reliability of Multiphysical Systems Set, Volume 9, Chapter 4 [DAH 21a]). Along these axes, the electric field propagates at different phase speeds. The proper axes are characterized by the so-called ordinary and extraordinary indices of the plate and which are determined by the properties of symmetry (see Reliability of Multiphysical Systems Set, Volume 9, Chapter 4 [DAH 21a]). Along these axes, the electric field propagates at different phase speeds. The displacement field D is parallel to E and its amplitude is proportional to that of E, such that $D_i = \varepsilon_i E_i$, where $\varepsilon_l = (n_i)^2$.

2) To which physical parameter do the terms fast and slow axes refer?

Solution:

The terms slow and fast axes correspond to the phase speeds given by $v_r = c/n_r$ and $v_l = c/n_l$, where n_i is the index along the axis. The vacuum index is 1 and the index in matter is greater than 1. The phase speed is therefore slower than that of light in vacuum. The slow term corresponds to the axis for which the phase speed is the lowest and conversely the fast term corresponds to the axis for which the phase speed is the highest.

3) Show that the axes introduce a phase difference $\Delta\varphi$ between the orthogonal components of the wave given by: $\Delta\varphi = \left({}^{2\pi}/_{\lambda}\right)(n_l - n_r)d$.

Solution:

On the slow axis, the phase of the wave is: $\Delta\varphi_l = \left({}^{2\pi}/_{\lambda}\right)n_l d$.

On the fast axis, the phase of the wave is: $\Delta\varphi_r = \left({}^{2\pi}/_{\lambda}\right)n_r d$.

The phase difference is: $\Delta\varphi = \Delta\varphi_l - \Delta\varphi_r = \left({}^{2\pi}/_{\lambda}\right)(n_l - n_r)d$.

4) A quarter-wave plate introduces a phase shift $\Delta\varphi = \pi/2$ and a half-wave plate introduces a phase shift $\Delta\varphi = \pi$ between the components of the electric field. Give the corresponding matrices of these plates.

Solution:

If the wave s is parallel to the axis Ox, Ox,

$$\vec{E}_s(\vec{r},t) = \left|\vec{E}_s\right|\exp\big(-i(\omega t)\big)\overrightarrow{e_x}.$$

If the wave p is parallel at axis Ox,

$$\vec{E}_p(\vec{r},t) = \left|\vec{E}_p\right|\exp\big(-i(\omega t)\big)\overrightarrow{e_y}.$$

After crossing the plate, the polarized wave has the expression $\vec{E}_s(\vec{r},t) = |\vec{E}_s| \exp\left(i\left(-\omega t + \left(2\pi/_\lambda\right)n_r d\right)\right)\overrightarrow{e_x}$ and $\vec{E}_p(\vec{r},t) = |\vec{E}_p| \exp\left(i\left(-\omega t + \left(2\pi/_\lambda\right)n_l d\right)\right)\overrightarrow{e_y}$. When the wave propagates along the z axis in the positive direction, the sign between the temporal and spatial phases is negative.

If the elements of the matrix *M* are denoted by *A*, *B*, *C* and *D*, such that $\begin{pmatrix} A & B \\ C & D \end{pmatrix}$, *B* and *C* are zero, so that the non-zero terms are $A = \exp\left(i\left(\left(2\pi/_\lambda\right)n_l d\right)\right) = \exp(i\Delta\varphi_l)$ and $D = \exp\left(i\left(\left(2\pi/_\lambda\right)n_r d\right)\right) = \exp(i\Delta\varphi_r)$. Then, the matrix can be expressed as:

$$\begin{pmatrix} e^{i\Delta\varphi_l} & 0 \\ 0 & e^{i\Delta\varphi_r} \end{pmatrix} = e^{\frac{i(\Delta\varphi_l+\Delta\varphi_r)}{2}} \begin{pmatrix} e^{\frac{i(\Delta\varphi_l-\Delta\varphi_r)}{2}} & 0 \\ 0 & e^{-\frac{i(\Delta\varphi_l-\Delta\varphi_r)}{2}} \end{pmatrix}$$

In the case of a quarter-wave plate, $\Delta nd = \lambda/4$ and a matrix *M* is obtained:

$$M = e^{\frac{i(\Delta\varphi_l+\Delta\varphi_r)}{2}} \begin{pmatrix} e^{\frac{i\pi}{4}} & 0 \\ 0 & e^{-\frac{i\pi}{4}} \end{pmatrix} = e^{i\psi} e^{\frac{i\pi}{4}} \begin{pmatrix} 1 & 0 \\ 0 & -i \end{pmatrix}$$

In the case of a half-wave plate, $\Delta nd = \lambda/2$, and a matrix *M* is obtained:

$$M = e^{\frac{i(\Delta\varphi_l+\Delta\varphi_r)}{2}} \begin{pmatrix} e^{\frac{i\pi}{2}} & 0 \\ 0 & e^{-\frac{i\pi}{2}} \end{pmatrix} = e^{i\psi} e^{\frac{i\pi}{2}} \begin{pmatrix} 1 & 0 \\ 0 & -1 \end{pmatrix}$$

As we generally calculate the intensity, we have $e^{i\psi} * e^{-i\psi} = 1$ and the matrices are:

$$M\left(\frac{\lambda}{4}\right) = e^{\frac{i\pi}{4}} \begin{pmatrix} 1 & 0 \\ 0 & -i \end{pmatrix} \text{ and } M\left(\frac{\lambda}{2}\right) = e^{\frac{i\pi}{2}} \begin{pmatrix} 1 & 0 \\ 0 & -1 \end{pmatrix}$$

1.4.3. *Interferometry and optical lambda meter*

In photonic systems, it is often necessary to precisely know the wavelength of the laser sources used by the systems. This is the case when the atmosphere is studied by lidars which use various lasers at various wavelengths or in telemetry for the measurement of distances. The Earth–Moon distance, for example, is measured from a pulsed laser (Côte d'Azur Observatory, Calern Plateau) which sends 10 pulses per second towards the Moon. The reflected photons are collected by a 1.5 m diameter telescope. The number of reflected photons is very low, of the order of one photon per 100 shots. The time interval between the emission of the light pulses and the reception of the return signal is between 2.3 and 2.8 seconds. This time interval is measured with an accuracy of 7–10 ps, which provides a distance between the transmitter and the receiver to within 3 mm on average.

Wavelength measuring systems are called lambda meters and are based on the use of a Michelson interferometer. The diagram of the device of a lambda meter is given in Figure 1.22. The main optical elements are a quarter-wave plate oriented at 45° (Q_{45}) with respect to the s axis, a polarizer *s* oriented at 45° with respect to the *s* axis (P_{45}) and a cube polarization splitter (CSP). A major difference compared to the conventional interferometer is that: in the case of a lambda meter, the two arms of the interferometer are mobile. This is done using a double-sided mirror, a mirror whose two sides are reflective (see Figure 1.22).

If $\vec{E}_S$ is the field of the light wave propagating on one of the arms of the interferometer and $\vec{E}_P$ the field of the wave propagating on the other arm, their expressions are given by:

$$\vec{E}_S(\vec{r},t) = \left|\vec{E}_S\right| \mathrm{e}^{i(-\omega t+\vec{k}\vec{r}_{0S})}\,\overrightarrow{e_S} = \left|\vec{E}_S\right| \mathrm{e}^{i(-\omega t+\varphi_{0S})}\,\overrightarrow{e_S} \qquad [1.11]$$

and

$$\vec{E}_P(\vec{r},t) = \left|\vec{E}_P\right| \mathrm{e}^{i(-\omega t+\vec{k}\vec{r}_{0P})}\,\overrightarrow{e_P} = \left|\vec{E}_P\right| \mathrm{e}^{i(-\omega t+\varphi_{0P})}\,\overrightarrow{e_P} \qquad [1.12]$$

where $\omega = 2\pi\nu_0$ is the pulse of the wave, ν_0 is the frequency of the wave equal to $\nu_0 = c_0/\lambda_0$ in a vacuum and $k = 2\pi/\lambda_0$ is the norm of the wave vector. $c_0 = 299\ 792\ 458\ \mathrm{ms}^{-1}$.

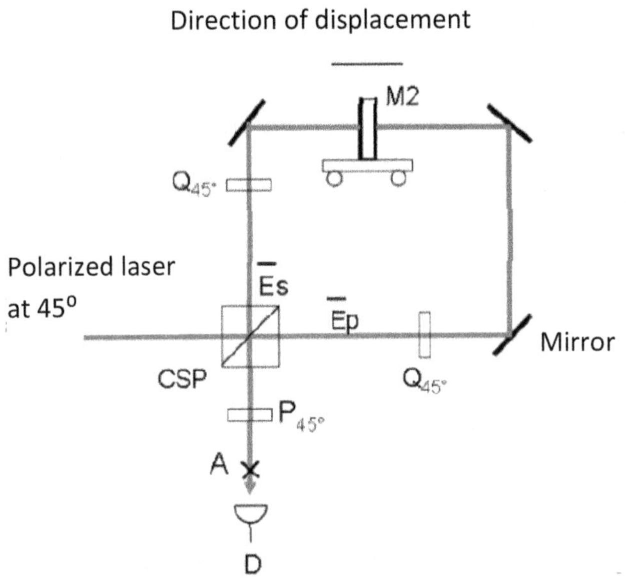

Figure 1.22. *Schematic diagram of the device of a lambda meter. For a color version of this figure, see www.iste.co.uk/dahoo/metrology2.zip*

In the Jones basis:

$$\vec{E}_L(\vec{r},t) = e^{-i\omega t}\begin{pmatrix}|\vec{E}_S|e^{i\varphi_{0S}}\\ |\vec{E}_P|e^{i\varphi_{0P}}\end{pmatrix} \quad [1.13]$$

If the laser wave is polarized at 45°, then $e^{i\varphi_{0S}} = e^{i\varphi_{0P}} = e^{i\varphi_0}$ and $|\vec{E}_S| = |\vec{E}_P|$.

To determine the expressions of the components of the resulting field at point A (Figure 1.20) after passing through the photonic device, the Jones matrix approach is used. The matrices of the different components are given in the appendix. r_s and r_p represent, respectively, the total distances traveled (source-round trip-detector) traveled by the *s* wave and the *p* wave between the polarization splitter cube and the mirror M_2.

1) On the outward path, the plate is Q_{45}, and on the return path, it is Q_{-45}. Why?

2) Which of these four statements corresponds to the mounting of the lambda meter (circle the correct answer)?

A: At the level of the cube CSP, if the electric field on the outward path is E_S (respectively, E_P), on the return path it becomes E_p ((respectively, E_S) after crossing the cube.

B: At the level of the CSP cube, if the electric field on the outward path is E_S (respectively, E_P), on the return path it becomes E_S ((respectively, E_P) after crossing the cube.

C: At the level of the CSP cube, if the electric field on the outward path is E_S (respectively, E_P), on the return path it becomes E_p ((respectively, E_P) after crossing the cube.

D: At the level of the CSP cube, if the electric field on the outward path is E_S (respectively, E_P), on the return path it becomes E_S ((respectively, E_S) after crossing the cube.

3) On the matrices associated with Q_{45} and Q_{-45} which are given by:

$$Q_{45} = \frac{1}{2}\begin{pmatrix} a & a^* \\ a^* & a \end{pmatrix} \text{ and } Q_{-45} = \frac{1}{2}\begin{pmatrix} a & -a^* \\ -a^* & a \end{pmatrix}$$

where $a = 1\text{-}i$ with $i^2 = -1$ and a^* is the complex conjugate of a: $a^* = 1+i$.

Check that $aa^* = 2$ and that $a^2+a^{*2} = 0$.

4) It will be assumed that P_{45} is defined on the return path. Show that:

$$P_{45} = \frac{1}{2}\begin{pmatrix} 1 & 1 \\ 1 & 1 \end{pmatrix}$$

5) Check that $M^2 = I$.

The initial laser field is given by: $\vec{E}_L = \vec{E}_S + \vec{E}_P = \begin{pmatrix} |E_S|e^{i\varphi_{os}} \\ |E_P|e^{i\varphi_{op}} \end{pmatrix} e^{-i\omega t}$ with:

$$\vec{E}_s(\vec{r},t) = |\vec{E}_s| \exp\left(i(-\omega t + \vec{k}\vec{r}_{0s})\right) \vec{e_s} = |\vec{E}_s| \exp(i(-\omega t + \varphi_{0s}))\vec{e_s}$$

and:

$$\vec{E}_p(\vec{r},t) = |\vec{E}_p| \exp\left(i(-\omega t + \vec{k}\vec{r}_{0p})\right) \vec{e_p} = |\vec{E}_p| \exp\left(i(-\omega t + \varphi_{0p})\right) \vec{e_p}$$

By using the formalism of the matrix equation for each path, we can calculate the electric field $E_S(A)$ arriving on the detector at A through the path S and the corresponding electric field $E_P(PD)$ on the path P. Like the laser is polarized at 45°, we have

6) From the initial laser field, we get on the path of the arm initially S, the field in A from the multiplication of the following matrices: $\vec{E}_S(A) = D(\varphi_S) * P_{45}^S * CSP_P * Q_{-45} * M_2 * Q_{45} * CSP_S * \vec{E}_L$. We do not have to consider the effect of the mirror between Q_{45} and M_2 which allows us to modify the path, because $M^2=I$, the identity matrix. Justify the matrices present in this product using the diagram of the device in Figure 1.22.

7) From the interpretation of the various Jones matrices: CSP_S, Q_{45}, M_2, Q_{-45}, CSP_P, $D(\varphi_s)$ and $P^S{}_{45}$ give the expanded matrix expressions.

8) By multiplying the matrices, determine the field: $\vec{E}_S(A) = -K_S \frac{|E_S| e^{i\varphi_{rS}}}{2} \begin{pmatrix}1\\1\end{pmatrix}$, where K_S, which includes the term $e^{i\omega t}$, is a complex constant and φ_{rS} is a phase shift. Give the expression of φ_{rS} as a function of n, λ_0 and r_s, the path traveled in the interferometer (use the expression of $\phi = 2\pi nd/\lambda_0$ given in the attached form, specifying d).

9) Calculate $E_P(A)$, the wave propagating on the arm P. By proceeding in the same way as for the field $E_S(A)$ give its expression in terms of $ABCD$ matrices.

10) Explain the matrices allowing us to calculate $EP(A)$ $E_P(A)$.

11) Show that: $\vec{E}_P(A) = K_P \frac{|E_P| e^{i\varphi_{rP}}}{2} \begin{pmatrix}1\\1\end{pmatrix}$, where K_P is a complex constant and φ_{rP} a phase shift. Express φ_{rp} as a function of n, λ_0 and r_p, the path traveled in the interferometer (use the expression of $\phi = 2\pi nd/\lambda_0$ given in the form in the appendix, specifying d).

12) Deduce the expression for the intensity of the light wave at the output of the interferometer when $r_s = r_p$, i.e. the two paths are identical in the two arms. Set $I_0 = |E_S|^2 + |E_P|^2 = 1$ and $I = \varepsilon_0 c <E>^2$.

13) The mobile mirror M_2 is moved by an amount Δz in the direction shown in Figure 1.22 to the right. Rewrite the previous expression as a function of Δz.

14) To which displacement (or variation of optical path) does a phase variation of 2π correspond?

15) This displacement corresponds to the scrolling of an interference fringe. Denote by N the number of interference fringes obtained for a displacement d. Give the expression of N as a function of n, d and λ_0. Calculate N for a displacement of 10 cm knowing that $\lambda_0 = 632.991458$ nm and $n = 1.000247$ (round to the nearest integer).

16) The previous laser is replaced by a laser whose wavelength is unknown. Let λ_i be this wavelength. For the same displacement, a number of fringes N_i is obtained. Give the expression of λ_i as a function only of λ_0, N and N_i. Deduce the value of λ_i knowing that $N_i = 512\ 643$. Provide the result with six significant digits after the decimal point.

17) It can be shown that the relative uncertainty on this measurement varies as $1/N$. What must then be the minimum displacement d_{min} corresponding to a relative uncertainty on λ_i of 10^{-6} (i.e. 0.001 nm). Calculate d_{min} within 1 mm.

1.4.3.1. *Answer to question 1*

On the outward path, the plate is Q_{45} and on the return path it is Q_{-45}. Why?

The axes of the path are locally defined so as to position the Oz axis on the light path and the Ox axis horizontally along the s polarization and the Oy axis vertically along the Oy axis parallel to the p polarization. On the figure given in appendix and in Figure 1.23, the go wave is in the direction Oz, which corresponds to a phase $\Psi = -\omega t + kr_s$ for the component s and a phase $\Psi = \omega t + kr_p$ for the p component for a phase velocity in the direction of increasing z.

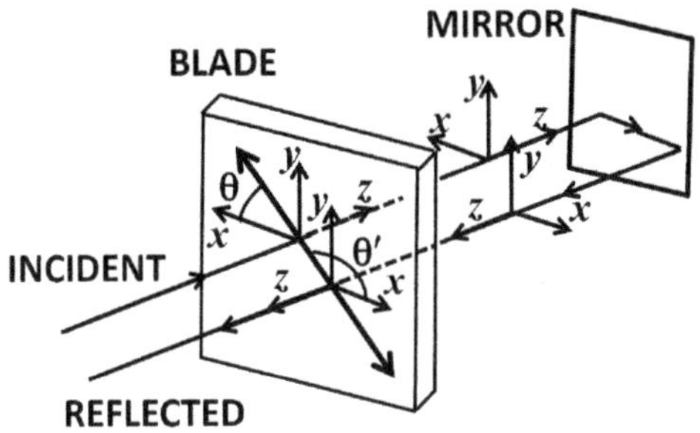

Figure 1.23. *Diagram of the light path and reference axes*

On the outward angle between the optical axis is 45° in the counterclockwise direction and on the return, the local axes are positioned differently on the plate and the angle made by the optical axis is 135° or -45° counterclockwise.

1.4.3.2. *Answer to question 2*

Which of these four statements corresponds to the mounting of the lambda meter (circle the correct answer)?

After reflection, the wave is in an s-type state and after transmission in a p-type state. The correct answer is given by Proposition A.

1.4.3.3. *Answer to question 3*

According to the form given in the appendix, aa*=$(1-i)(1+i) = 1-(-1) = 2$.

Thus: a^2+a^{*2} =$(1-i)^2$+$(1+i)^2 = -2i + 2i = 0$.

1.4.3.4. *Answer to question 4*

Using the formulas:

$M(\theta) = R(\theta)M(0)R(-\theta)$ and $R(\theta) = \begin{pmatrix} \cos\theta & -\sin\theta \\ \sin\theta & \cos\theta \end{pmatrix}$ with θ = 45°,

we calculate:

$$P(45°) = \frac{1}{\sqrt{2}}\begin{pmatrix} 1 & -1 \\ 1 & 1 \end{pmatrix}\begin{pmatrix} 1 & 0 \\ 0 & 0 \end{pmatrix}\frac{1}{\sqrt{2}}\begin{pmatrix} 1 & 1 \\ -1 & 1 \end{pmatrix} = \frac{1}{2}\begin{pmatrix} 1 & -1 \\ 1 & 1 \end{pmatrix}\begin{pmatrix} 1 & 1 \\ 0 & 0 \end{pmatrix} = \frac{1}{2}\begin{pmatrix} 1 & 1 \\ 1 & 1 \end{pmatrix}$$

1.4.3.5. *Answer to question 5*

$$M * M = \begin{pmatrix} 1 & 0 \\ 0 & -1 \end{pmatrix} \begin{pmatrix} 1 & 0 \\ 0 & -1 \end{pmatrix} = \begin{pmatrix} 1 & 0 \\ 0 & 1 \end{pmatrix}$$

1.4.3.6. *Answer to question 6*

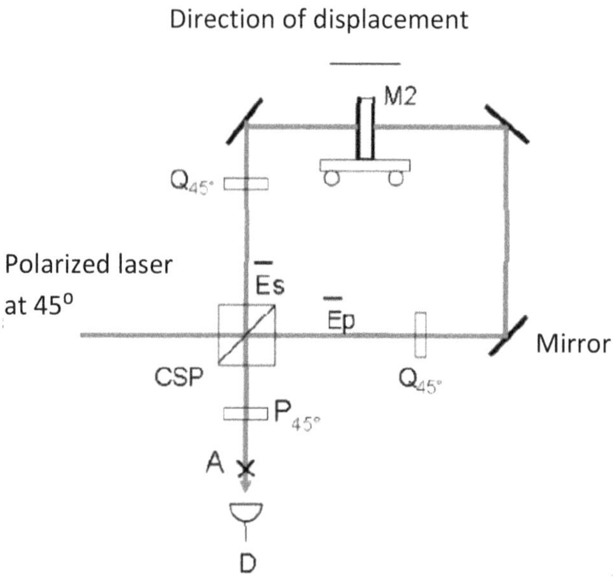

Reminder: Figure 1.22. *Schematic diagram of the device of a lambda meter*

In Figure 1.22, the matrices are in order: *CSPS*, which reflects the component *s* of the incident field, then the quarter-wave plate Q^{45}, then a first mirror, then the mirror M2, the mirror again (2 times reflection, M2 = 1), the quarter-wave plate on the return path, i.e. $Q\text{-}^{45}$, transmission in the cube, therefore transformation into a *p* wave, hence *CSPp*, and the polarizer *s* at 45° before the detector.

To these transformations by the optical elements should be added the path of the beam retaining the *s* part of the wave, hence the multiplication by the matrix *D*(φ*s*) which switches with all the other matrices.

1.4.3.7. *Answer to question 7*

$$CSP_S = \begin{pmatrix} 1 & 0 \\ 0 & 0 \end{pmatrix} \quad Q_{45} = \frac{1}{2}\begin{pmatrix} 1+i & 1-i \\ 1-i & 1+i \end{pmatrix} \quad M_2 = \begin{pmatrix} 1 & 0 \\ 0 & -1 \end{pmatrix}$$

$$Q_{-45} = \frac{1}{2}\begin{pmatrix} 1+i & -(1-i) \\ -(1-i) & 1+i \end{pmatrix} \quad CSP_p = \begin{pmatrix} 0 & 0 \\ 0 & 1 \end{pmatrix} \quad P_{S45} = \frac{1}{2}\begin{pmatrix} 1 & 1 \\ 1 & 1 \end{pmatrix}$$

$$D(\varphi) = \begin{pmatrix} \exp(i\varphi) & 0 \\ 0 & \exp(i\varphi) \end{pmatrix}$$

$\varphi = nk_0\delta = 2\pi n\delta/\lambda_0$ with $\delta = r_{is}$ the round trip between the source and the detector, traversed by the light on the arm where the cube returns by reflection the polarization s.

Under MATLAB (appendix): >> $M_S = P^S{}_{45}*CSP_P*Q_{-45}*M2*Q45*CSPS$;
leads to the matrix $M = -\frac{1}{2}\begin{pmatrix} 1 & 0 \\ 1 & 0 \end{pmatrix}$.

Multiplication by $D(\varphi)$ leads to $-\frac{1}{2}\begin{pmatrix} \exp(i\varphi) & 0 \\ \exp(i\varphi) & 0 \end{pmatrix}$

1.4.3.8. *Answer to question 8*

$$\vec{E}_S(A) = -\frac{1}{2}\begin{pmatrix} \exp(i\varphi) & 0 \\ \exp(i\varphi) & 0 \end{pmatrix} * \vec{E}_L = -\frac{1}{2}\begin{pmatrix} \exp(i\varphi) & 0 \\ \exp(i\varphi) & 0 \end{pmatrix}\begin{pmatrix} |E_S|e^{i\varphi_{os}} \\ |E_P|e^{i\varphi_{op}} \end{pmatrix} e^{-i\omega t}$$

$$\vec{E}_S(A) = -\frac{1}{2}e^{i\varphi}\begin{pmatrix} |E_S|e^{i\varphi_{os}} \\ |E_S|e^{i\varphi_{os}} \end{pmatrix} e^{-i\omega t} = -K_S\frac{|E_S|e^{i\varphi_{rs}}}{2}\begin{pmatrix} 1 \\ 1 \end{pmatrix}$$

By identification: a $e^{i\varphi} = e^{i\varphi_{rs}}$ and $K_S = e^{i\varphi_{os}}e^{-i\omega t}$

The expression of $\varphi_{rS} = 2\pi n(r_s)/\lambda_0$.

1.4.3.9. *Answer to question 9*

On the other arm, we have:

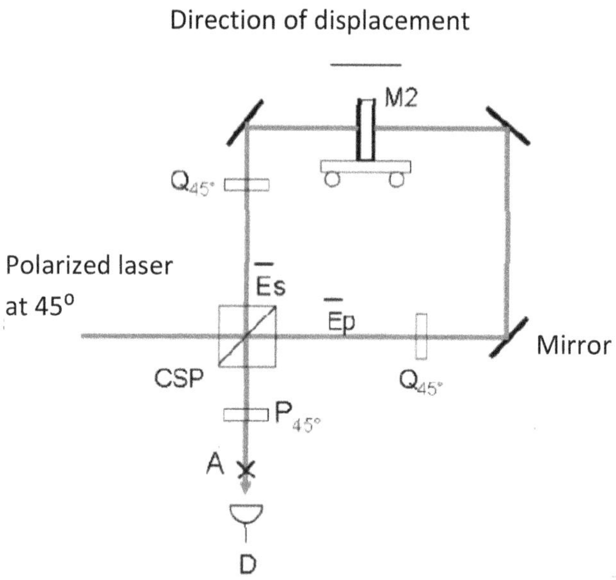

Reminder: Figure 1.22. *Schematic diagram of the device of a lambda meter*

A path by transmission in the cube, either *CSPP*, then the crossing of the quarter-wave plate, or *Q*45, the reflection on the mirror once on the outward path and once on the return, which amounts to multiplying by *I*, of even for the following mirror, the reflection on M2, or M2, the crossing of the quarter-wave plate on the return path or Q^{-45}, the reflection on the cube corner, or *CSPS*, then the crossing of the polarizer at 45° and the propagation matrix $D(\varphi_p)$ on the arm *p* by transmission.

Thus: $\vec{E}_P(A) = D(\varphi_P) * P_{45}^S * CSP_S * Q_{-45} * M_2 * Q_{45} * CSP_P * \vec{E}_L$.

1.4.3.10. *Answer to question 10*

From the matrices given in question 7 only $D(\varphi)$ differs:

$$D(\varphi) = \begin{pmatrix} \exp(i\varphi) & 0 \\ 0 & \exp(i\varphi) \end{pmatrix}$$

where $\varphi = nk_0\delta = 2\pi n\delta/\lambda_0$ with $\delta = r_{ip}$ the round trip between the source and the detector, traveled by light on the arm where the cube operates in transmission and returns the polarization *p*.

1.4.3.11. *Answer to question 11*

The multiplication of matrices leads to:

$$\vec{E}_P(A) = \frac{1}{2}\begin{pmatrix} 0 & \exp(i\varphi) \\ 0 & \exp(i\varphi) \end{pmatrix} * \vec{E}_L = \frac{1}{2}\begin{pmatrix} 0 & \exp(i\varphi) \\ 0 & \exp(i\varphi) \end{pmatrix}\begin{pmatrix} |E_S|e^{i\varphi_{0S}} \\ |E_P|e^{i\varphi_{0P}} \end{pmatrix} e^{-i\omega t}$$

$$\vec{E}_P(A) = \frac{1}{2}e^{i\varphi}\begin{pmatrix} |E_P|e^{i\varphi_{0P}} \\ |E_P|e^{i\varphi_{0P}} \end{pmatrix} e^{-i\omega t} = K_P\frac{|E_P|e^{i\varphi_{rP}}}{2}\begin{pmatrix} 1 \\ 1 \end{pmatrix}$$

It can be obtained by identification: a $e^{i\varphi} = e^{i\varphi_{rP}}$ and $K_P = e^{i\varphi_{0P}}e^{-i\omega t}$

and $\varphi_{rP}=2\pi n(r_P)/\lambda_0$.

1.4.3.12. *Answer to question 12*

$$\frac{I}{\varepsilon_0 c} = \left|\vec{E}_S(A) + \vec{E}_P(A)\right|^2 = \left(\vec{E}_S(A) + \vec{E}_p(A)\right)^* \left(\vec{E}_S(A) + \vec{E}_p(A)\right)$$

where:

$$\frac{I}{\varepsilon_0 c} = \vec{E}_S(A)\left(\vec{E}_S(A)\right)^* + \vec{E}_p(A)\left(\vec{E}_p(A)\right)^* + \vec{E}_S(A)\left(\vec{E}_p(A)\right)^* + \vec{E}_p(A)\left(\vec{E}_S(A)\right)^*$$

Since: $\vec{E}_S(A) = -K_S\frac{|E_S|e^{i\varphi_{rs}}}{2}\begin{pmatrix} 1 \\ 1 \end{pmatrix}$ and $\vec{E}_P(A) = K_P\frac{|E_P|e^{i\varphi_{rP}}}{2}\begin{pmatrix} 1 \\ 1 \end{pmatrix}$, then:

$$\vec{E}_S(A)\left(\vec{E}_S(A)\right)^* = (K_S)^*\frac{|E_S|e^{-i\varphi_{rs}}}{2}\begin{pmatrix} 1 \\ 1 \end{pmatrix}K_S\frac{|E_S|e^{i\varphi_{rs}}}{2}\begin{pmatrix} 1 \\ 1 \end{pmatrix}$$
$$= |K_S|^2\frac{|E_S|^2}{4}\begin{pmatrix} 1 \\ 1 \end{pmatrix}.\begin{pmatrix} 1 \\ 1 \end{pmatrix}\left|\vec{E}_S(A)\right|^2 = |K_S|^2\frac{|E_S|^2}{4}2 = |K_S|^2\frac{|E_S|^2}{2} \text{ and}$$
$$\left|\vec{E}_P(A)\right|^2 = |K_P|^2\frac{|E_P|^2}{4}2 = |K_P|^2\frac{|E_P|^2}{2}$$

For the cross term:

$$\vec{E}_S(A)\left(\vec{E}_p(A)\right)^* = -K_S\frac{|E_S|e^{i\varphi_{rs}}}{2}\begin{pmatrix} 1 \\ 1 \end{pmatrix}.(K_P)^*\frac{|E_P|e^{-i\varphi_{rP}}}{2}\begin{pmatrix} 1 \\ 1 \end{pmatrix}$$

$$\vec{E}_S(A)\left(\vec{E}_p(A)\right)^* = -K_S(K_P)^*\frac{|E_S||E_P|e^{i(\varphi_{rs}-\varphi_{rP})}}{2}$$

$$\vec{E}_P(A)\left(\vec{E}_S(A)\right)^* = -K_P(K_S)^* \frac{|E_S||E_P|e^{-i(\varphi_{rS}-\varphi_{rP})}}{2}$$

Given the expression of the initial laser field:

$$\vec{E}_L = \vec{E}_S + \vec{E}_P = \begin{pmatrix} |E_S|e^{i\varphi_{0S}} \\ |E_P|e^{i\varphi_{0P}} \end{pmatrix} e^{-i\omega t}$$

then: $\frac{I_L}{\varepsilon_0 c} = \vec{E}_L\left(\vec{E}_L\right)^* = |E_S|^2 + |E_P|^2 = I_0$ and in this case:

$$cos\alpha = \frac{|E_S|}{\sqrt{(|E_S|^2+|E_P|^2)}} \text{ and } sin\alpha = \frac{|E_P|}{\sqrt{(|E_S|^2+|E_P|^2)}}.$$

As $K_S = e^{i\varphi_{0S}}e^{-i\omega t}$ and $K_P = e^{i\varphi_{0P}}e^{-i\omega t}$; using the fact that at the output of the laser, the wave is polarized at 45°, then $e^{i\varphi_{0S}} = e^{i\varphi_{0P}}$, i.e. $K_S = K_P = K_0$, and $|E_S|^2 = |E_P|^2 = 1/2$, thus:

$$\frac{I}{\varepsilon_0 c} = |K_0|^2\frac{|E_S|^2}{2} + |K_0|^2\frac{|E_P|^2}{2} - |K_0|^2|E_S||E_P|\frac{e^{i(\varphi_{rS}-\varphi_{rP})} + e^{-i(\varphi_{rS}-\varphi_{rP})}}{2}$$

$$\frac{I}{\varepsilon_0 c} = \frac{|K_0|^2}{2}(|E_S|^2 + |E_P|^2)(1 - \frac{2|E_S||E_P|}{(|E_S|^2 + |E_P|^2)}\cos(\varphi_{rS} - \varphi_{rP}))$$

and since $\varphi_{rS} - \varphi_{rP} = \frac{2\pi n}{\lambda}(r_S - r_P)$:

$$\frac{I}{\varepsilon_0 c} = \frac{|K_0|^2}{2}(1 - \cos(\frac{2\pi n}{\lambda}(r_S - r_P))$$

Since $r_S - r_P = 0$:

$$\frac{I}{\varepsilon_0 c} = \frac{|K_0|^2}{2}(1 - 1) = 0$$

1.4.3.13. *Answer to question 13*

If the mirror is moved to the right, on path r_S, the path is $2\Delta z$ longer, and on path r_P, the path is $2\Delta z$ shorter. Then:

$$\frac{I}{\varepsilon_0 c} = \frac{|K_0|^2}{2}(1 - \cos(\frac{2\pi n}{\lambda}(r_S + 2\Delta z - (r_P - 2\Delta z))))$$

$$\frac{I}{\varepsilon_0 c} = \frac{|K_0|^2}{2}(1 - \cos(\frac{2\pi n}{\lambda}(r_s - r_P + 4\Delta z)))$$

$$\frac{I}{\varepsilon_0 c} = \frac{|K_0|^2}{2}(1 - \cos(\frac{2\pi n}{\lambda}4\Delta z))$$

1.4.3.14. *Answer to question 14*

$$\frac{2\pi n}{\lambda}4\Delta z = 2\pi \Rightarrow \Delta z = \frac{\lambda}{4n}$$

1.4.3.15. *Answer to question 15*

$$N = \frac{d}{\Delta z} \Rightarrow N = 4n\frac{d}{\lambda}$$

Digital application: N = 4x1.000247x10 10^{-2}/(632.991458 10^{-9}) = 632 076

1.4.3.16. *Answer to question 16*

$$\Delta z = \frac{\lambda}{4n} \text{ and } N = \frac{d}{\Delta z} \Rightarrow N = 4n\frac{d}{\lambda}$$

The ratio is thus: $\frac{N_i}{N} = \frac{4n\frac{d}{\lambda_i}}{4n\frac{d}{\lambda}} \Rightarrow \lambda_i = \lambda\frac{N}{N_i}$

Digital application:

λ_I = 632.991458$10^{-9}$ × (632 076/512 643) = 780.462639 10^{-9} m

1.4.3.17. *Answer to question 17*

$$N_i = 4n\frac{d}{\lambda_i}$$

If the relative uncertainty is equal to 10^{-6}, then N is equal to 10^6. In this case, d is given by:

$d = \frac{N_i\lambda_i}{4n}$ = 10^6× 780.462639 10^{-9}/(4×1.000247) = 19.5 mm.

1.4.4. *The homodyne interferometer and refractometer*

The electric field is usually composed of a component E_p which is parallel to the plane of incidence and of a component E_s perpendicular to the plane of incidence. The coordinate axes for the *p* wave and *s* waves are given in Figure 1.24.

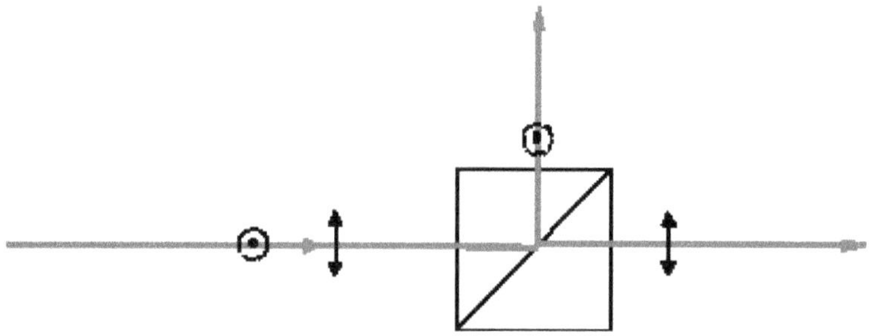

Figure 1.24. *Coordinate axes for p wave and s wave. For a color version of this figure, see www.iste.co.uk/dahoo/metrology2.zip*

The refractive index of a liquid medium can be measured using a laser interferometer. The sample to be analyzed is placed in one of the arms of the interferometer, and the refractive index of the medium is obtained by comparing the phase difference between the two arms of the measuring instrument.

Homodyne interferometers are commonly used in quantum optics [GRO 01, WEN 05, FUW 15].

A homodyne interferometer is represented schematically in Figure 1.25. The laser used is a helium–neon laser emitting an unpolarized light beam at a wavelength of 632.991 nm in vacuum. The instrument is equipped with an S polarizer (P^{45}) oriented at 45°, a quarter-wave plate oriented at 45° (- Q^{45}) and a polarization splitter cube (CSP) which is called CSPS (reflection mode) or CSPS (transmission mode).

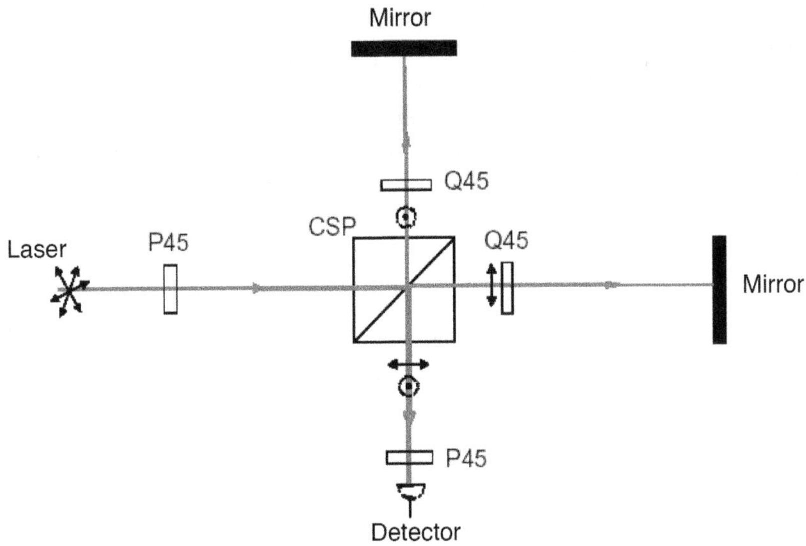

Figure 1.25. *Diagram of a homodyne interferometer and of the paths s and p. For a color version of this figure, see www.iste.co.uk/dahoo/metrology2.zip*

Let E_s be the field of the light wave propagating on one of the interferometer arms, and E_p, the field of the wave propagating on the other arm such that:

$$\vec{E}_s(\vec{r},t) = |\vec{E}_s| \exp(i\vec{k}\vec{r_0}) \exp\left(i(-\omega t + \vec{k}\vec{r_s})\right) \vec{e_s}$$

and

$$\vec{E}_p(\vec{r},t) = |\vec{E}_p| \exp(i\vec{k}\vec{r_0}) \exp\left(i(-\omega t + \vec{k}\vec{r_p})\right) \vec{e_p}$$

where $\omega = 2\pi\nu_0$ is the pulse of the wave, and ν_0 the frequency of the wave being equal to $\nu_0 = c/\lambda_0$ in a vacuum. We give $c = 299\ 792\ 458\ \mathrm{ms}^{-1}$.

To obtain the resulting electric field at the photodetector apply Jones formalism, and use the vectors and matrices associated with the optical elements and the rotation matrices, the reference defined by the polarization s and p of the incident wave (see the Appendix).

1) Give the expression of the components of the resulting field at the level of the photodetector. This result will be obtained applying the Jones matrix calculation. The matrices of the different components are given in the appendix. r_s and r_p represent, respectively, the total distances traveled by the s wave and by the p wave between the laser source after the polarizer and the photodetector on the corresponding arms.

2) Obtain the expression of the intensity of the light wave at the output of the interferometer. Set $I_0 = |E_S|^2 + |E_P|^2 = 1$ and $I = \varepsilon_0 c < E >^2$.

3) In order to measure the refractive index n_L of the liquid medium, a sample containing this liquid is placed in the fixed arm of the interferometer (Figure 1.26).

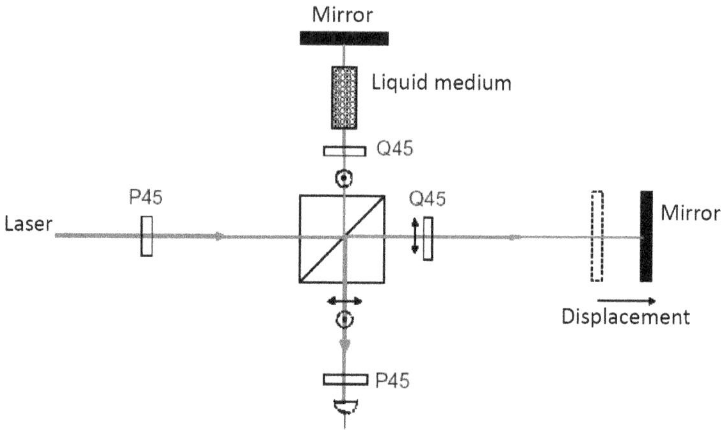

Figure 1.26. *Diagram of the interferometric measurement device. For a color version of this figure, see www.iste.co.uk/dahoo/metrology2.zip*

Since the length of the cell containing the liquid medium is L and its index is n_L, obtain from question 2 the expression of the light intensity. The mobile mirror is fixed and it is taken that $r_s = r_p$.

4) The moving mirror is displaced by a quantity Δz in the direction indicated in Figure 1.26 in order to compensate for the phase shift

induced by the introduction of the liquid medium and measure n_L. Express the light intensity as a function of Δz.

5) A displacement of 13.35 mm is measured. Since n_{air} = 1.000274 and L = 10.32 mm, calculate the refractive index of the liquid medium.

1.4.4.1. *Answer to question 1*

From the paths made by the two components *s* and *p*, the component *s* of the electric field on the photodetector is expressed by:

$$\vec{E}_s(PD) = D(\varphi_S) * P_{45}^S * CSPP * Q_{-45} * M * Q_{45} * CSPS * P_{45} * \vec{E}_L$$

The component *p* on the photodetector is expressed by:

$$\vec{E}_P(PD) = D(\varphi_P) * P_{45}^S * CSPS * Q_{-45} * M * Q_{45} * CSPP * P_{45} * \vec{E}_L$$

with: $\vec{E}_L = \begin{pmatrix} |\vec{E}_0| e^{ikr_0} e^{-i\omega t} \\ |\vec{E}_0| e^{ikr_0} e^{-i\omega t} \end{pmatrix}$

P_{45} (*S*), Q_{45}, Q_{-45} and *CSP* are then calculated:

$$P_{45}^S = \frac{1}{2}\begin{pmatrix} 1 & 1 \\ 1 & 1 \end{pmatrix},\ Q_{45} = \frac{1}{2}\begin{pmatrix} 1+i & 1-i \\ 1-i & 1+i \end{pmatrix},\ Q_{-45} = \frac{1}{2}\begin{pmatrix} 1+i & -1+i \\ -1+i & 1+i \end{pmatrix}$$

$$CSP_S = \begin{pmatrix} 1 & 0 \\ 0 & 0 \end{pmatrix} \quad CSP_P = \begin{pmatrix} 0 & 0 \\ 0 & 1 \end{pmatrix} \quad M = \begin{pmatrix} 1 & 0 \\ 0 & -1 \end{pmatrix}$$

$$D(\varphi_S) = \begin{pmatrix} \exp(i\varphi_S) & 0 \\ 0 & \exp(i\varphi_S) \end{pmatrix} \text{ and } D(\varphi_P) = \begin{pmatrix} \exp(i\varphi_P) & 0 \\ 0 & \exp(i\varphi_P) \end{pmatrix}$$

With $\varphi_S = nkr_S = 2\pi nr_S/\lambda$ and $\varphi_P = nkr_P = 2\pi nr_P/\lambda$, where r_S and r_P correspond to the round trips between the source and the detector on each arm.

Multiplying the matrices leads to:

$$\vec{E}_s(PD) = -\frac{1}{2}\begin{pmatrix} e^{inkr_s} & 0 \\ e^{inkr_s} & 0 \end{pmatrix} * \vec{E}_L = -\frac{1}{2}|\vec{E}_0| e^{ikr_0} e^{-i\omega t} e^{inkr_s} \begin{pmatrix} 1 \\ 1 \end{pmatrix}$$

and

$$\vec{E}_p(PD) = \frac{1}{2}\begin{pmatrix} 0 & e^{inkr_P} \\ 0 & e^{inkr_P} \end{pmatrix} * \vec{E}_L = \frac{1}{2}\left|\vec{E}_0\right| e^{ikr_0} e^{-i\omega t} e^{inkr_P} \begin{pmatrix} 1 \\ 1 \end{pmatrix}$$

where: $\vec{E}_L(PD) = \vec{E}_S(PD) + \vec{E}_P(PD) = \frac{1}{2}\begin{pmatrix} -e^{inkr_s} & e^{inkr_P} \\ -e^{inkr_s} & e^{inkr_P} \end{pmatrix} * \vec{E}_L$

$$\vec{E}_L(PD) = -\frac{1}{2}\left|\vec{E}_0\right| e^{ikr_0} e^{-i\omega t} e^{inkr_s} \begin{pmatrix} 1 \\ 1 \end{pmatrix} + \frac{1}{2}\left|\vec{E}_0\right| e^{ikr_0} e^{-i\omega t} e^{inkr_P} \begin{pmatrix} 1 \\ 1 \end{pmatrix}$$

$$\vec{E}_L(PD) = -\frac{1}{2}\left|\vec{E}_0\right| e^{ikr_0} e^{-i\omega t} e^{inkr_s} \begin{pmatrix} 1 \\ 1 \end{pmatrix} + \frac{1}{2}\left|\vec{E}_0\right| e^{ikr_0} e^{-i\omega t} e^{inkr_P} \begin{pmatrix} 1 \\ 1 \end{pmatrix}$$

$$\vec{E}_L(PD) = \frac{1}{2}\left|\vec{E}_0\right| e^{ikr_0} e^{-i\omega t} \begin{pmatrix} -e^{inkr_s} + e^{inkr_P} \\ -e^{inkr_s} + e^{inkr_P} \end{pmatrix}$$

1.4.4.2. *Answer to question 2*

As $I = \varepsilon_0 c_0 \langle \vec{E}_L(PD).\vec{E}_L(PD)^* \rangle$

$$\frac{I}{\frac{\varepsilon_0 c_0}{2}} = \left|\vec{E}_S(PD)\right|^2 + \left|\vec{E}_P(PD)\right|^2 + 2(\vec{E}_S(PD)\vec{E}_P{}^*(PD) + \vec{E}_P(PD)\vec{E}_S{}^*(PD)$$

and: $\left|\vec{E}_S(PD)\right|^2 = \frac{1}{2}\left|\vec{E}_0\right|^2$

and $\left|\vec{E}_P(PD)\right|^2 = \frac{1}{2}\left|\vec{E}_0\right|^2$

and $\left|\vec{E}_S(PD)\right|^2 + \left|\vec{E}_P(PD)\right|^2 = \left|\vec{E}_0\right|^2$

The crossed terms lead to:

$$\vec{E}_S(PD)\vec{E}_P{}^*(PD) + \vec{E}_P(PD)\vec{E}_S{}^*(PD) = -\frac{1}{2}\left|\vec{E}_0\right|^2 (e^{in(kr_s - kr_P)} + e^{-in(kr_s - kr_P)})$$

$$\frac{I}{\frac{\varepsilon_0 c_0}{2}} = \left|\vec{E}_0\right|^2 - \left|\vec{E}_0\right|^2 \cos(kr_s - kr_P) = \left|\vec{E}_0\right|^2 \left(1 - \cos(\frac{2\pi}{\lambda} n(r_s - r_P)\right)$$

Since $I_0 = \left|\vec{E}_0\right|^2 = 1$

$$\frac{I}{\frac{\varepsilon_0 c_0}{2}} = \left(1 - \cos(\frac{2\pi}{\lambda} n(r_s - r_P)\right)$$

This can be written using the matrix form by:

$$\vec{E}_L(PD) = \vec{E}_S(PD) + \vec{E}_P(PD) = \frac{1}{2}\begin{pmatrix} -e^{ikr_s} & e^{ikr_P} \\ -e^{ikr_s} & e^{ikr_P} \end{pmatrix} * \vec{E}_L$$

$$\frac{I}{\frac{\varepsilon_0 c_0}{2}} = \vec{E}_L(PD).\vec{E}_L(PD)^*$$

$$\frac{I}{\frac{\varepsilon_0 c_0}{2}} = \begin{pmatrix} -e^{ikr_s} & e^{ikr_P} \\ -e^{ikr_s} & e^{ikr_P} \end{pmatrix}\begin{pmatrix} \left|\vec{E}_0\right| e^{ikr_0} e^{-i\omega t} \\ \left|\vec{E}_0\right| e^{ikr_0} e^{-i\omega t} \end{pmatrix}.\begin{pmatrix} -e^{-ikr_s} & e^{-ikr_P} \\ -e^{-ikr_s} & e^{-ikr_P} \end{pmatrix}\begin{pmatrix} \left|\vec{E}_0\right| e^{-ikr_0} e^{i\omega t} \\ \left|\vec{E}_0\right| e^{-ikr_0} e^{i\omega t} \end{pmatrix}$$

Finally: $I_F(PD) = \frac{\varepsilon_0 c_0}{2}\left(1 - \cos(\frac{2\pi n}{\lambda}(r_p - r_s)\right)$

1.4.4.3. *Answer to question 3*

Since $r_p = r_s$, the only path difference between the two paths in each arm comes from the passage of the light beam in the liquid of index n_L, in the arm connected to the fixed mirror where the polarization is *s*. Consequently:

$$I_F(PD) = \frac{\varepsilon_0 c_0}{2}\left(1 - \cos(\frac{2\pi}{\lambda}(\left(r_p - (r_s - 2L + 2n_L L)\right)\right)$$

$$I_F(PD) = \frac{\varepsilon_0 c_0}{2}\left(1 - \cos(\frac{2\pi}{\lambda}(2(n_L - 1)L)\right)$$

1.4.4.4. *Answer to question 4*

If the mobile mirror moves to Δz, there is a differential path length on the arm connected to the mobile mirror, hence:

$$I_F(PD) = \frac{\varepsilon_0 c_0}{2}\left(1 - \cos(\frac{2\pi}{\lambda}(2n_{air}\Delta z - 2(n_L - 1)L)\right)$$

1.4.4.5. *Answer to question 5*

The intensity is zero at the detector when the cosine is equal to 1, hence the following equation is obtained:

$$2n_{air}\Delta z - 2(n_L - 1)L = 0 \Rightarrow n_L - 1 = \frac{n_{air}\Delta z}{L}.$$

Consequently: $n_L = 1 + \frac{n_{air}\Delta z}{L} = 1 + \frac{1{,}000274 \times 10{,}05}{10{,}32} = 1{,}974\ 104$

1.4.5. *The heterodyne interferometer*

In the diagram of the Michelson interferometer in Figure 1.27, a laser beam is split into two beams by a non-polarizing 50/50 beam splitter plate. The arm where the reflection takes place on the moving mirror is the measurement beam and the other arm where the beam is reflected on the fixed mirror is the reference beam. On the return optical path, the two beams cross the splitter plate. A part of each beam returns to the laser source, while the rest of the beams are superimposed on the photodiode. The intensity on the photodiode is directly related to the displacement δ of the mobile mirror. In this configuration, the displacement information is a direct current (d.c.). However, d.c. signals are difficult to deal with in the low frequency domain because of multiple disturbances such as, for example, measurement time drifts caused by amplifier gains. It is much easier to obtain and analyze the displacement information in the form of an alternative current (a.c.) signal. To generate this a.c. signal, a heterodyne interferometer design based on two lasers of frequencies ω_1 and ω_2 is used.

The principle of the heterodyne interferometer is based on the Doppler effect. A heterodyne interferometer has two beams of frequencies ω_1 and ω_2 close to a few megahertz, which are linearly and orthogonally polarized with respect to each other and are spatially separated by a polarization splitter cube.

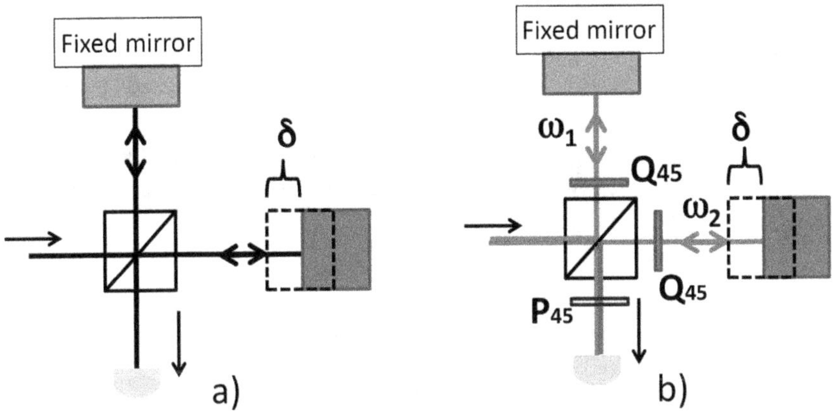

Figure 1.27. *Laser interferometer: a) homodyne and b) heterodyne. For a color version of this figure, see www.iste.co.uk/dahoo/metrology2.zip*

The behavior of the reference beam is similar to that of the homodyne interferometer. Initially, it is s polarized. Then, it is reflected by the beam splitter plate in the direction of the reference mirror. After a double passage through a quarter-wave plate, it passes through the separator plate again to be mixed with the measurement beam. The measurement beam is reflected by the moving mirror moving at a constant speed v, which has the effect of shifting its frequency by Δv_D by the Doppler effect. At the output of the interferometer, the two beams, measure and reference, are mixed using a polarizer generating an interference pattern. After passing through a polarizer (45°), the two beams are superimposed on a photodiode. The information on the displacement of the witness mirror is in phase with the sinusoidal signal at the frequency $\omega = |\omega_1 - \omega_2|$. The beat frequency between the two beams, observed at the photodetector, thus constitutes the measurement signal S_3.

The fields of application are dimensional metrology, microelectronics and production chains for high-tech industries. Integrated circuits are made by photolithography. Photolithography is a process that reproduces a pattern, specific to the desired chip, by projection of light onto a silicon substrate (or wafer) of a mechanical mask. A semiconductor component is made from several layers layered on top of each other. At each layer, the wafer is removed from the device for treatment and then put back into place for the projection of the next layer. In order to ensure the interconnection of integrated circuits on the previous layer, the wafer must be repositioned to the nanometer scale (10^{-9} m). For example, in 256 MB Flash memories, the distance between two lines is 80 nm. Decreasing this distance amounts to increasing the capacity of the component or its speed. The economic stakes involved in wanting to have positioning systems that are as exact as possible are important. These systems are monitored in manufacturing plants by Michelson interferometers (Figure 1.28).

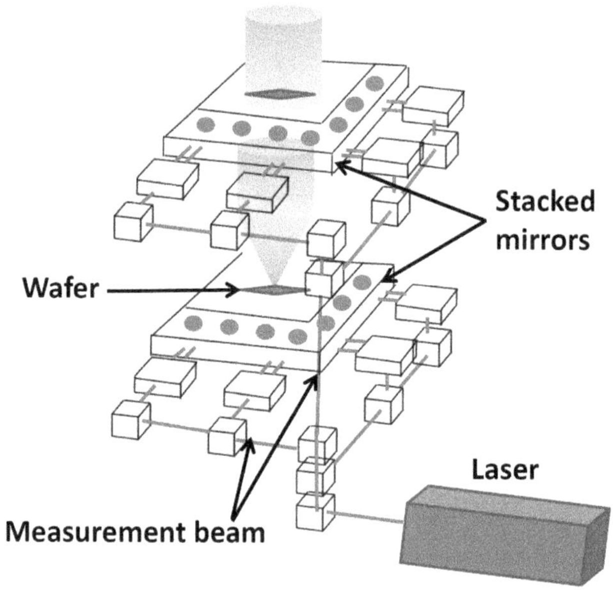

Figure 1.28. *Dimensional metrology. For a color version of this figure, see www.iste.co.uk/dahoo/metrology2.zip*

1.4.5.1. *Exercise: heterodyne interferometer*

The heterodyne interferometer [SUT 87, OLD 93, DEM 98, TOP 03, SCH 06] uses a laser source emitting two monochromatic plane waves of the same amplitude and the same initial phase but with crossed polarizations. The angular frequencies (or pulsations) of the two beams are different. ω_1 is the pulsation of the *s* wave and ω_2 that of the *p* wave. The principle of operation is as follows: when the moving mirror moves, the frequency of the light wave is shifted due to the Doppler effect. By measuring the Doppler frequency, it is possible to deduce the speed of the mirror (RADAR principle but at radiometric frequencies). Knowing the speed, the displacement by means of a time measurement is obtained.

The following Figure 1.29 displays the optical device design and the elements of a heterodyne interferometer with cube corners as reflectors.

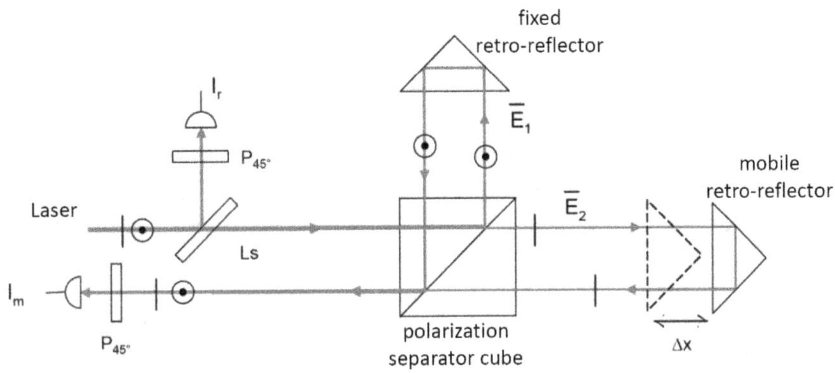

Figure 1.29. *Heterodyne laser interferometer with cube corners. For a color version of this figure, see www.iste.co.uk/dahoo/metrology2.zip*

Several modern technologies use the Doppler effect. Radar, for example (RAdio Detection And Ranging), was developed just before World War II to detect and locate enemy airplanes at long distances. Radar pointers are based on the Doppler effect. Such direct read pointers are used on sports fields to measure ball speed. The laser source emits a monochromatic beam stabilized in frequency which passes through an acousto-optical modulator (AOM) which generates two beams polarized at 90°, of wavelengths λ_1 = 632.991528 nm and λ_2 = 632.991501 nm at different frequencies [DEM 98, TOP 03]. The frequency difference is 20 MHz.

Using the Lorentz transformation formulas, it can be deduced in spectroscopy [DAH 19] that a molecule moving with a speed $\vec{v}$, in the laboratory frame of reference, absorbs radiation of wave number $\omega = 2\pi c/\lambda$ given by:

$$\omega = \omega_0 \left(1 + \frac{\vec{v}.\vec{k}}{c}\right) \qquad [1.14]$$

where ω_0 is the laser probe wave number corresponding to the transition between the two energy levels of the molecule, $\vec{k}$ is a unit vector in the direction of the beam and c is the speed of light.

Likewise, if ω_0 is the frequency of the light wave arriving on a mobile mirror moving at a speed V, after reflection on the mirror, the reflected beam frequency is $\omega_d = \omega_0(1 \pm \frac{2nV}{c})$, where c = 299 792 458 ms^{-1} is the speed of light in vacuum and n = 1.000247 is the refractive index of air. The sign ± depends on the direction of movement. If the mirror gets closer to (away from) the source, then the + (-). sign will be used. The speed of the mirror V can then be deduced.

1) Calculate using matrix calculus, the expressions for the intensities I_{ref} and I_{mes}. Set $|\vec{E}_s| = |\vec{E}_p| = 1$, $\Delta\omega = \omega_2 - \omega_1$ and neglect the initial phase terms.

2) The associated electronics circuit measures the phase difference and uses this difference to provide the position and speed as measurement data. Calculate the Doppler frequency associated with the displacement of the mobile mirror at a speed of 2 mm.s^{-1}. Since the detector bandwidth is 12 GHz, calculate the limit speed measurable by this method if the associated electronics circuit has a 26.6 MHz bandwidth.

3) The travel time is measured with an ultra-stable Quartz clock. Deduce from (1) the displacement Δx.

4) What should be the time measurement uncertainty to obtain an uncertainty of 1 nm on the displacement.

1.4.5.1.1. Solution to question 1

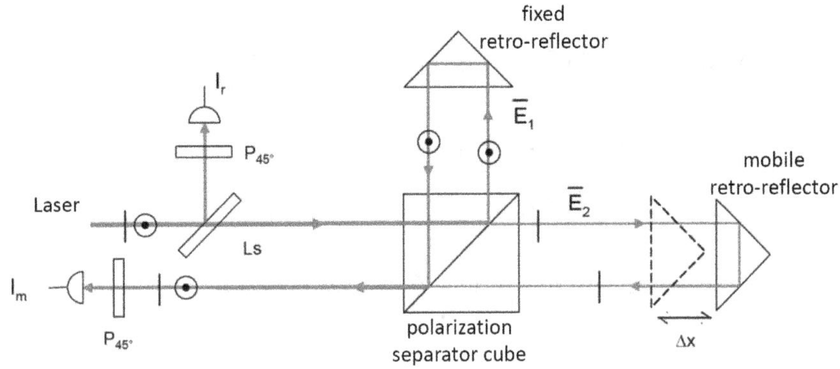

Reminder: Figure 1.29. *Heterodyne laser interferometer with cube corners*

The reference beam

The reference beam is detected before it passes through the interferometer by a detector while crossing a 45° polarizer (P_{45}).

Consequently:

$$\vec{E}_L = \begin{pmatrix} |\vec{E}_0| e^{ikr_0} e^{-i\omega_1 t} \\ |\vec{E}_0| e^{ikr_0} e^{-i\omega_2 t} \end{pmatrix} = \begin{pmatrix} |\vec{E}_0| e^{ikr_0} e^{-i\omega_1 t} \\ 0 \end{pmatrix} + \begin{pmatrix} 0 \\ |\vec{E}_0| e^{ikr_0} e^{-i\omega_2 t} \end{pmatrix}$$

From the paths of the two components *s* and *p*, the component *s* of the electric field on the photodetector is:

$$\vec{E}_s(PD) = D(\varphi_S) * P_{45}^S * \vec{E}_S$$

and that the component *p* of the electric field on the photodetector is:

$$\vec{E}_P(PD) = D(\varphi_P) * P_{45}^S * \vec{E}_P$$

Since $P_{45}^S = \frac{1}{2}\begin{pmatrix} 1 & 1 \\ 1 & 1 \end{pmatrix}$ and $D(\varphi) = \begin{pmatrix} e^{i\varphi} & 0 \\ 0 & e^{i\varphi} \end{pmatrix}$

$$\vec{E}_s(PD) = \begin{pmatrix} e^{i\varphi_S} & 0 \\ 0 & e^{i\varphi_S} \end{pmatrix} \frac{1}{2}\begin{pmatrix} 1 & 1 \\ 1 & 1 \end{pmatrix} \begin{pmatrix} |\vec{E}_0| e^{ikr_0} e^{-i\omega_1 t} \\ 0 \end{pmatrix} = \frac{1}{2}\begin{pmatrix} |\vec{E}_0| e^{i\varphi_S} e^{ikr_0} e^{-i\omega_1 t} \\ |\vec{E}_0| e^{i\varphi_S} e^{ikr_0} e^{-i\omega_1 t} \end{pmatrix}$$

and

$$\vec{E}_P(PD) = \begin{pmatrix} e^{i\varphi_P} & 0 \\ 0 & e^{i\varphi_P} \end{pmatrix} \frac{1}{2} \begin{pmatrix} 1 & 1 \\ 1 & 1 \end{pmatrix} \begin{pmatrix} 0 \\ |\vec{E}_0| e^{ikr_0} e^{-i\omega_1 t} \end{pmatrix} = \frac{1}{2} \begin{pmatrix} |\vec{E}_0| e^{i\varphi_P} e^{ikr_0} e^{-i\omega_2 t} \\ |\vec{E}_0| e^{i\varphi_P} e^{ikr_0} e^{-i\omega_2 t} \end{pmatrix}$$

The intensity calculation is obtained using the relation:

$$I = \varepsilon_0 c_0 \vec{E}_L(PD).\vec{E}_L(PD)^*$$

$$\frac{I}{\varepsilon_0 c_0} = |\vec{E}_S(PD)|^2 + |\vec{E}_P(PD)|^2 + 2(\vec{E}_S(PD)\vec{E}_P^{\,*}(PD) + \vec{E}_P(PD)\vec{E}_S^{\,*}(PD)$$

Since: $|\vec{E}_S(PD)|^2 = \frac{1}{2}|\vec{E}_0|^2$ and $|\vec{E}_P(PD)|^2 = \frac{1}{2}|\vec{E}_0|^2$

and: $|\vec{E}_S(PD)|^2 + |\vec{E}_P(PD)|^2 = |\vec{E}_0|^2$

Considering equal initial phases, the crossed terms lead to:

$$\vec{E}_S(PD)\vec{E}_P^{\,*}(PD) + \vec{E}_P(PD)\vec{E}_S^{\,*}(PD) = \frac{1}{2}|\vec{E}_0|^2(e^{-i(\omega_1-\omega_2)t} + e^{i(\omega_1-\omega_2)t})$$

$$\vec{E}_S(PD)\vec{E}_P^{\,*}(PD) + \vec{E}_P(PD)\vec{E}_S^{\,*}(PD) = \frac{1}{2}|\vec{E}_0|^2(e^{i(\Delta\omega)t} + e^{-i\Delta\omega)t})$$

$$\frac{I}{\varepsilon_0 c_0} = |\vec{E}_0|^2 - |\vec{E}_0|^2 \cos(\Delta\omega t) = |\vec{E}_0|^2(1 - \cos(\Delta\omega t))$$

Since $|\vec{E}_0|^2 = 1$, the calculation of the intensity leads to: $\frac{I}{\varepsilon_0 c_0} = (1 - \cos(\Delta\omega t))$.

The measurement beam

The measurement beam travels through the interferometer.

The S-type measurement beam is reflected by the polarization splitter cube (*CSPS*), then on two cube corners ($M^2 = 1$), before being reflected again on the polarization splitter cube (*CSPS*) and crosses a 45° polarizer (P_{45}) before reaching the detector (P_{45} * *CSPS* * *CSPS*).

The P-type measurement beam passes through the polarization splitter cube (*CSPP*), then it is reflected on two cube corners ($M^2 = 1$) moving *D*(*d*), before crossing the splitter cube again polarization (CSPP) and finally passes through a 45° polarizer (P_{45}) before reaching the detector (P_{45} * *CSPP* * *D*(*d*) * *CSPP*).

Analyzing the paths made by the two components *s* and *p*, the component s of the electric field on the photodetector is obtained:

$$\vec{E}_s(PD) = D(\varphi_S) * P_{45}^S * CSPS ** CSPS * \vec{E}_S$$

and the component *p* of the electric field on the photodetector is:

$$\vec{E}_P(PD) = D(\varphi_P) * P_{45}^S * CSPP * D(d) * CSPP * \vec{E}_P$$

Since:

$$\vec{E}_L = \begin{pmatrix} |\vec{E}_0| e^{ikr_0} e^{-i\omega_1 t} \\ |\vec{E}_0| e^{ikr_0} e^{-i\omega_2 t} \end{pmatrix} = \begin{pmatrix} |\vec{E}_0| e^{ikr_0} e^{-i\omega_1 t} \\ 0 \end{pmatrix} + \begin{pmatrix} 0 \\ |\vec{E}_0| e^{ikr_0} e^{-i\omega_2 t} \end{pmatrix}$$

Using the P_{45} (*S*), *CSPS* and *CSPP* matrices, the phase shift matrix *D*(φ) and the Doppler effect matrix *D*(*d*), it is obtained that:

$$P_{45}^S = \frac{1}{2}\begin{pmatrix} 1 & 1 \\ 1 & 1 \end{pmatrix}, \; CSP_S = \begin{pmatrix} 1 & 0 \\ 0 & 0 \end{pmatrix} \quad CSP_P = \begin{pmatrix} 0 & 0 \\ 0 & 1 \end{pmatrix} \quad M = \begin{pmatrix} 1 & 0 \\ 0 & -1 \end{pmatrix}$$

$$D(\varphi_S) = \begin{pmatrix} \exp(i\varphi_S) & 0 \\ 0 & \exp(i\varphi_S) \end{pmatrix} \text{ and } D(\varphi_P) = \begin{pmatrix} \exp(i\varphi_P) & 0 \\ 0 & \exp(i\varphi_P) \end{pmatrix}$$

$$D(d) = \begin{pmatrix} 0 & 0 \\ 0 & \exp(-i\omega_D t) \end{pmatrix}$$

with $\varphi_S = nkr_S = 2\pi nr_S/\lambda$ and $\varphi_P = nkr_P = 2\pi nr_P/\lambda 1$ where r_S and r_P correspond to the paths between the source and the detector on each arm, the multiplication of matrices leads to:

$$\vec{E}_s(PD) = \frac{1}{2}\begin{pmatrix} e^{inkr_S} & 0 \\ e^{inkr_S} & 0 \end{pmatrix} * \vec{E}_L = \frac{1}{2}|\vec{E}_0| e^{ikr_0} e^{-i\omega_1 t} e^{inkr_S} \begin{pmatrix} 1 \\ 1 \end{pmatrix}$$

and

$$\vec{E}_P(PD) = \frac{1}{2}\begin{pmatrix} 0 & e^{inkr_P} \\ 0 & e^{inkr_P} \end{pmatrix} * \begin{pmatrix} 0 & 0 \\ 0 & e^{\pm i\omega_D t} \end{pmatrix} * \vec{E}_L = \frac{1}{2}\left|\vec{E}_0\right| e^{ikr_0} e^{-i(\omega_2 \pm \omega_D)t} e^{inkr_P} \begin{pmatrix} 1 \\ 1 \end{pmatrix}$$

From where:

$$\vec{E}_L(PD) = \vec{E}_S(PD) + \vec{E}_P(PD) = \frac{1}{2}\begin{pmatrix} e^{inkr_S} & e^{\pm i\omega_D t} e^{inkr_P} \\ e^{inkr_S} & e^{\pm i\omega_D t} e^{inkr_P} \end{pmatrix} * \vec{E}_L$$

$$\vec{E}_L(PD) = \frac{1}{2}\left|\vec{E}_0\right| e^{ikr_0} e^{-i\omega_1 t} e^{inkr_S} \begin{pmatrix} 1 \\ 1 \end{pmatrix} + \frac{1}{2}\left|\vec{E}_0\right| e^{ikr_0} e^{-i\omega_2 t} e^{\pm i\omega_D t} e^{inkr_P} \begin{pmatrix} 1 \\ 1 \end{pmatrix}$$

The intensity calculation is made from: $I = \varepsilon_0 c_0 \vec{E}_L(PD).\vec{E}_L(PD)^*$

$$\frac{I}{\varepsilon_0 c_0} = \left|\vec{E}_S(PD)\right|^2 + \left|\vec{E}_P(PD)\right|^2 + 2(\vec{E}_S(PD)\vec{E}_P^*(PD) + \vec{E}_P(PD)\vec{E}_S^*(PD)$$

Since:

$$\left|\vec{E}_S(PD)\right|^2 = \tfrac{1}{2}\left|\vec{E}_0\right|^2 \text{ and } \left|\vec{E}_P(PD)\right|^2 = \tfrac{1}{2}\left|\vec{E}_0\right|^2$$

and by hypothesis $\left|\vec{E}_S(PD)\right|^2 + \left|\vec{E}_P(PD)\right|^2 = \left|\vec{E}_0\right|^2$, the crossed terms lead to:

$$\vec{E}_S(PD)\vec{E}_P^*(PD) + \vec{E}_P(PD)\vec{E}_S^*(PD) = \frac{1}{2}\left|\vec{E}_0\right|^2 (e^{-i(\omega_1 - \omega_2 \pm \omega_D)t} + e^{i(\omega_1 - \omega_2 \pm \omega_D)t})$$

$$\vec{E}_S(PD)\vec{E}_P^*(PD) + \vec{E}_P(PD)\vec{E}_S^*(PD) = \frac{1}{2}\left|\vec{E}_0\right|^2 (e^{i(\Delta\omega \mp \omega_D)t} + e^{-i\Delta\omega \mp \omega_D)t})$$

$$\frac{I}{\varepsilon_0 c_0} = \left|\vec{E}_0\right|^2 - \left|\vec{E}_0\right|^2 \cos(\,(\Delta\omega \mp \omega_D)t) = \left|\vec{E}_0\right|^2 (1 - \cos((\Delta\omega \mp \omega_D)t))$$

Since $\left|\vec{E}_0\right|^2 = 1$,

$$\frac{I}{\varepsilon_0 c_0} = (1 - \cos((\Delta\omega \mp \omega_D)t))$$

Note that the photodetectors, which detect the intensities, carry out the multiplication of the signals and deliver electrical signals at frequencies $\Delta\omega$ and $\Delta\omega \mp \omega_D$, where $\omega_D = \omega_2 \frac{2nV}{c}$.

1.4.5.1.2. Answer to question 2

The laser source emits a monochromatic beam of wavelength λ = 632.991501 nm in vacuum. After reflection on the mirror, the pulse of the wave becomes $\omega_d = \omega_2(1 \pm \frac{2nV}{c})$, where c = 299 792 458 ms^{-1} is the speed of light in vacuum and n = 1.000247 is the air refractive index.

In this case: $\omega_D = \omega_2 \frac{2nV}{c} = \frac{2\pi c}{\lambda}\frac{2nV}{c}$

and: $\omega_2 = \frac{2\pi c}{\lambda} = \frac{2\pi \times 299\ 792\ 458}{632.991501\ 10^{-9}} = 2.975792825\ 10^{15} rad/s$

$$\frac{2nV}{c} = \frac{2 \times 1.000247 \times 210^{-3}}{299\ 792\ 458} = 1.33459\ 10^{-11}$$

$$\omega_D = \omega_2 \frac{2nV}{c} = 2.975792825\ 10^{15} \times 1.33459\ 10^{-11} = 3.9714\ 10^{4} rad/s$$

Expressed as a frequency, $f_D = 6320.76 Hz = 6.32 kHz$

In the pulse space, the reference signal is observed at the frequency $\Delta\omega$ and the measuring signal at the pulse $\Delta\omega + \omega_D$.

Since λ_1 = 632.991528 nm and λ_2 = 632.991501 nm,

$$\Delta\omega = \frac{2\pi c}{\lambda_2} - \frac{2\pi c}{\lambda_1} = 2\pi \times 299\ 792\ 458 \left(\frac{1}{632.991501\ 10^{-9}} - \frac{1}{632.991528\ 10^{-9}}\right) = 1.26931251\ 10^{8} \frac{rad}{s}.$$

In frequency, this corresponds to: Δf = 20.201736 MHz. The spectrum of the signal in the frequency space is shown in Figure 1.30.

Since f the bandwidth is 26.6 MHz, the maximum signal is 13.3 MHz to the right or left of the frequency Δf.

In frequency, this corresponds to: $f_D = 13.3\ MHz$; in maximum displacement speed: $V = \frac{c f_D}{2 n f_2} = \frac{\lambda_2 f_D}{2n} = \frac{632.991501\ 10^{-9} \times 13{,}3 10^{6}}{2 \times 1.000247} = 4.2 m/s$.

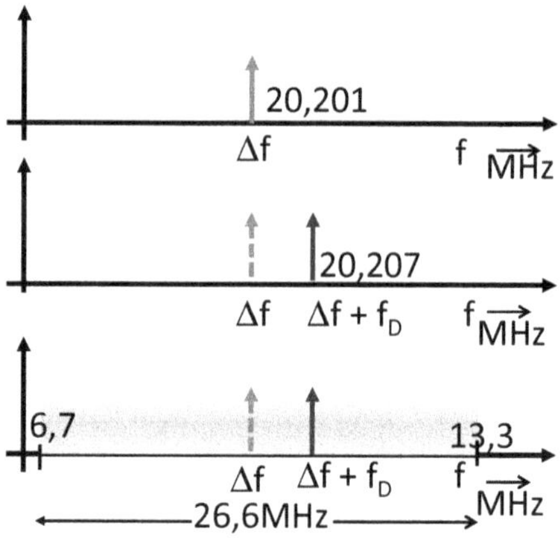

Figure 1.30. *Electrical signals measured in the detection bandwidth. For a color version of this figure, see www.iste.co.uk/dahoo/metrology2.zip*

Using a double-pass interferometer on the measuring path, a speed of 2.1 m/s can be measured. Note that a speed measurement of 1.9 km/s could be performed if the bandwidth was 12 GHz.

1.4.5.1.3. Answer to question 3

The speed is given by the displacement per unit of time. Consequently:

$$\Delta x = V\Delta t$$

The displacement at instant t is calculated by: $x(t) = \int_0^t \frac{\omega_D(t')}{2n\omega_2} c dt'$

If the measurement data are digital to improve resolution and precision, the displacement at time t is obtained by:

$$x(t) = \frac{c}{4\pi n\omega_2} \sum_{i=1}^{N} \Delta\Phi_i, \text{ where } N = t/\Delta T.$$

1.4.5.1.4. Answer to question 4

Resoluton is limited by the system noise and the phase measurement method [OLD 93]. The conventional digital phase meter detection techniques are mainly classified into three types: phase lock, fringe counting and zero crossing. A phase meter usually measures the rising time duration of the measurement signal relative to the counting clock, which is often set at 40 MHz. The phase resolution determines the position resolution. The output values of the phase meter are encoded on 8 bits. The incremental value, which is a fraction of the period of the clock at 40 MHz, for example, corresponds to the measurement of the rise time duration of the measurement signal. The phase resolution, which is 1 over 512 (9 bits because the state is 0 or 1), *h*, gives the position resolution of $\lambda/512\ M$, where *M* is the number of passes of the beam through the interferometer. For $M = 1$, we have $\lambda/512=1$ nm.

Regarding the measurement time, the equation, $\Delta t = \Delta x / V$, makes it possible to relate the uncertainty on *t* to the uncertainty on *x*.

1.4.6. *Application exercises on ellipsometry*

Ellipsometry is a non-destructive photonic technique, sensitive to surfaces and interfaces and which characterizes from a structural and optical point of view thin, mono or multilayer layers and massive materials (Chapter 8 of [DAH 16]). It is widely used in the microelectronics industry to control the thickness of "wafers". It is also used to monitor in real time the growth of a thin film or a multilayer stack (Reliability of Multiphysical Systems Set, Volume 9, Chapter 1 [DAH 21a]).

This optical analysis technique is based on the study of the polarization state of polarized light after it is reflected from a surface. In most cases, the state of polarization is elliptical (Figure 1.31) (Chapter 3 of [DAH 16]). By determining the characteristics of this elliptical state, the thicknesses and optical indices of the materials under study can be determined, hence the name ellipsometry (Ψ, Δ).

An ellipsometer consists of optical devices on two arms and a sample holder (Chapter 8, [DAH 16]). One of the arms has a light source and a set of optical devices for obtaining from the source a wave in a known state of polarization. The second arm includes an analyzer to determine the state of polarization of the wave after its reflection on the sample is characterized. A quarter-wave plate or a more elaborate optical device is generally used as a compensator (*C*), either on the polarizer side (*P*) (PCSA mount) or on the analyzer side (*A*) (PSCA mount) to cancel the effect of the sample (*S*) on the state of polarization of the reflected light. Figure 1.32 displays the diagram of a phase modulation ellipsometer which includes a birefringent modulator whose optical axes can be modulated by an alternating electric voltage, thus modulating the state of polarization of the light passing through it. The polarizer or the analyzer can be rotated to vary the polarization state of light over time. A detection synchronous with the modulation frequency of the polarized light is performed to extract the signal corresponding to the light which probes the material under study.

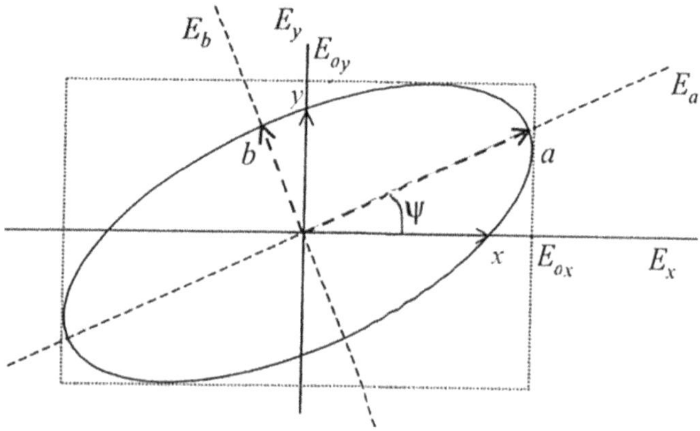

Figure 1.31. *Elliptical polarization state and ellipsometric parameters [DAH 16]*

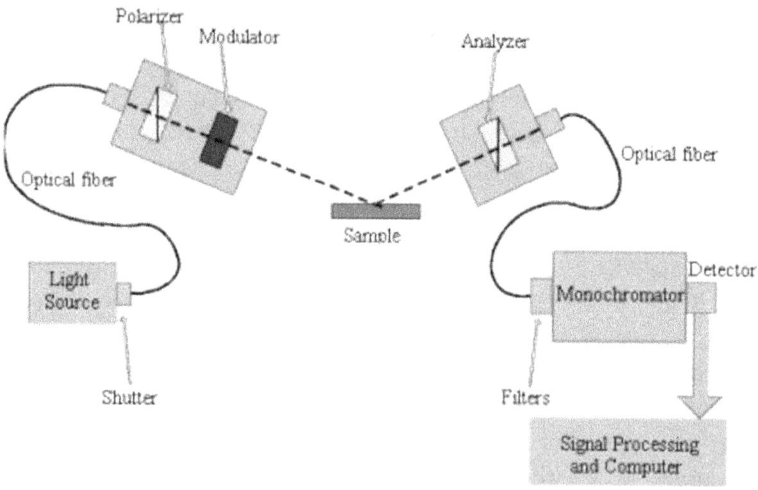

Figure 1.32. *Diagram of a phase modulation ellipsometer. For a color version of this figure, see www.iste.co.uk/dahoo/metrology2.zip*

1.4.6.1. *Exercise: ellisometry*

The photonic system presented in Figure 1.33 is in RAE mode (Rotating Analyzer Ellipsometer). The source is a helium–neon laser emitting a light beam with a wavelength in vacuum of 632, 791 nm. S designates the sample that interacts with light, and P^s (45°) is an S polarizer whose optical axis is oriented at 45° with respect to the axes of the frame of reference (*S*, *P*). P^s (*A*) is an *S* polarizer whose optical axis rotates at a variable angle *A* with respect to the axes of the frame of reference (*S*, *P*).

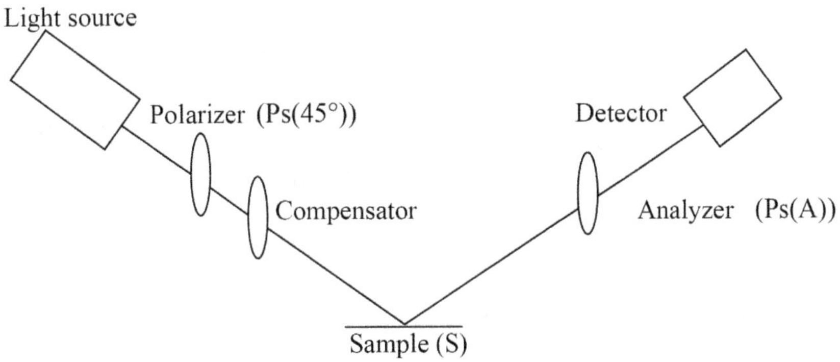

Figure 1.33. *Diagram of an ellipsometer in PCSA mode and reflected light. For a color version of this figure, see www.iste.co.uk/dahoo/metrology2.zip*

1) Formally, write the Jones equation that governs this photonic system.

Solution:

Since:

$$\vec{E}_L = \begin{pmatrix} |\vec{E}_0| e^{ikr_0} e^{-i\omega_1 t} \\ |\vec{E}_0| e^{ikr_0} e^{-i\omega_2 t} \end{pmatrix} = \begin{pmatrix} |\vec{E}_0| e^{ikr_0} e^{-i\omega_1 t} \\ 0 \end{pmatrix} + \begin{pmatrix} 0 \\ |\vec{E}_0| e^{ikr_0} e^{-i\omega_2 t} \end{pmatrix}$$

P_{45} (*S*), *C*, the effect of the sample *SE*, P^S (*A*), the phase shift matrix $D(\varphi)$ are used to calculate the electric field at the detector:

$$\vec{E}_L(PD) = D(\varphi) * P_{45}^A * SE * C * P_{45}^S * \vec{E}_L$$

and: $P_{45}^{S} = \frac{1}{2}\begin{pmatrix}1 & 1\\1 & 1\end{pmatrix}$, $C = \begin{pmatrix}e^{i\varphi_1} & 0\\0 & e^{i\varphi_2}\end{pmatrix}$, $SE = \begin{pmatrix}\sin(\Psi)\exp(i\nabla) & 0\\0 & \cos(\Psi)\end{pmatrix}$,

$$P_{45}^{A} = \begin{pmatrix}\cos^2 A & sinAcosA\\ sinAcosA & \sin^2 A\end{pmatrix},\ D(\varphi) = \begin{pmatrix}\exp(i\varphi) & 0\\0 & \exp(i\varphi)\end{pmatrix}$$

$\varphi_I = nkr_i = 2\pi nr_i/\lambda$, where r_i corresponds to the path between the source and the detector in the area concerned.

2) Using Jones' matrix calculation and assuming that an S-type polarizer is placed in front of the detector show that the E_A laser field at the output of the device is of the following form:

$$\vec{E}_A(PD) = |\vec{E}_0|e^{-i\omega t}\begin{pmatrix} sinAcosA\cos(\Psi) + \cos^2 A\sin(\Psi)\exp(i\nabla)e^{i\varphi} \\ \sin^2 A\cos(\Psi)\,e^{i\varphi_2} + sinAcosA\sin(\Psi)\exp(i\nabla)e^{i\varphi}\end{pmatrix}$$

Solution:

After crossing the polarizer oriented at 45°, the electric field is:

$$\vec{E}_L(P_{45}) = \begin{pmatrix}e^{i\varphi} & 0\\0 & e^{i\varphi}\end{pmatrix}\frac{1}{2}\begin{pmatrix}1 & 1\\1 & 1\end{pmatrix}\begin{pmatrix}|\vec{E}_0|e^{ikr_0}e^{-i\omega t}\\|\vec{E}_0|e^{ikr_0}e^{-i\omega t}\end{pmatrix} = \begin{pmatrix}|\vec{E}_0|e^{i\varphi}e^{ikr_0}e^{-i\omega t}\\|\vec{E}_0|e^{i\varphi}e^{ikr_0}e^{-i\omega t}\end{pmatrix}$$

It is a circularly polarized wave.

After the compensator:

$$\vec{E}_C(PD) = \begin{pmatrix}e^{i\varphi_1} & 0\\0 & e^{i\varphi_2}\end{pmatrix}\begin{pmatrix}|\vec{E}_0|e^{-i\omega t}\\|\vec{E}_0|e^{-i\omega t}\end{pmatrix} = \begin{pmatrix}|\vec{E}_0|e^{i\varphi_1}e^{-i\omega t}\\|\vec{E}_0|e^{i\varphi_2}e^{-i\omega t}\end{pmatrix}$$

where we reset the phase corresponding to the displacement.

After the sample:

$$\vec{E}_{SE}(PD) = \begin{pmatrix}\sin(\Psi)\exp(i\nabla) & 0\\0 & \cos(\Psi)\end{pmatrix}\begin{pmatrix}|\vec{E}_0|e^{i\varphi_1}e^{-i\omega t}\\|\vec{E}_0|e^{i\varphi_2}e^{-i\omega t}\end{pmatrix}$$
$$= \begin{pmatrix}|\vec{E}_0|\sin(\Psi)\exp(i\nabla)e^{i\varphi_1}e^{-i\omega t}\\|\vec{E}_0|\cos(\Psi)\,e^{i\varphi_2}e^{-i\omega t}\end{pmatrix}$$

At the level of the analyzer, the matrix corresponding to the 45° polarizer rotating through an angle A must be calculated:

$$P_{45}^{A} = \begin{pmatrix} \cos^2 A & sinAcosA \\ sinAcosA & \sin^2 A \end{pmatrix}$$

$$\vec{E}_P(PD) = \begin{pmatrix} \cos^2 A & sinAcosA \\ sinAcosA & \sin^2 A \end{pmatrix} \begin{pmatrix} |\vec{E}_0| \sin(\Psi) \exp(i\nabla) e^{i\varphi_1} e^{-i\omega t} \\ |\vec{E}_0| \cos(\Psi)\, e^{i\varphi_2} e^{-i\omega t} \end{pmatrix}$$

$$\vec{E}_L(PD) = |\vec{E}_0| e^{-i\omega t} \begin{pmatrix} sinAcosA\cos(\Psi)\, e^{i\varphi_2} + \cos^2 A \sin(\Psi) \exp(i\nabla) e^{i\varphi_1} \\ \sin^2 A \cos(\Psi)\, e^{i\varphi_2} + sinAcosA \sin(\Psi) \exp(i\nabla) e^{i\varphi_1} \end{pmatrix}$$

thus:

$$\vec{E}_L(PD) = |\vec{E}_0| e^{-i\omega t} e^{i\varphi_2} \begin{pmatrix} sinAcosA\cos(\Psi) + \cos^2 A \sin(\Psi) \exp(i\nabla) e^{i\varphi} \\ \sin^2 A \cos(\Psi) + sinAcosA \sin(\Psi) \exp(i\nabla) e^{i\varphi} \end{pmatrix}$$

3) Deduce that the expression of the intensity of the light wave IA at the level of the detector is of the form:

$$\frac{I}{\varepsilon_0 c_0} = |\vec{E}_0|^2 (\alpha + \beta cos2A + \gamma sin2A + \eta cos4A + \cdots)$$

Solution:

The intensity calculation is made from: $I = \varepsilon_0 c_0 \vec{E}_A(PD).\vec{E}_A(PD)^*$

$$\frac{I}{\varepsilon_0 c_0} = |\vec{E}_A(PD)|^2$$

$$\begin{aligned} |\vec{E}_S(PD)|^2 = |\vec{E}_0|^2 &\times \left(sinAcosA\cos(\Psi) + \cos^2 A \sin(\Psi) \exp(i\nabla) e^{i\varphi}\right) \\ &\times \left(sinAcosA\cos(\Psi) + \cos^2 A \sin(\Psi) \exp(-i\nabla)\, e^{-i\varphi}\right) \\ &+ \ (\sin^2 A \cos(\Psi) + sinAcosA \sin(\Psi) \exp(i\nabla)\, e^{i\varphi} \\ &\times (\sin^2 A \cos(\Psi) + sinAcosA \sin(\Psi) \exp(-i\nabla) e^{-i\varphi}) \end{aligned}$$

where:

$$\left|\vec{E}_S(PD)\right|^2 = \left|\vec{E}_0\right|^2\Big((sin^2Acos^2Acos^2(\Psi)) + (\cos^4Asin^2(\Psi)) \\ + 2\,sinAcosAcos(\Psi)\cos^2Asin(\Psi)\cos(\nabla+\varphi)\Big) \\ + \Big((\sin^4Acos^2(\Psi)) + (sin^2Acos^2Asin^2(\Psi))\Big) \\ + 2\,sinAcosAsin(\Psi)\sin^2Acos(\Psi)\cos(\nabla+\varphi)$$

Using the trigonometric formulas given in the appendix, we can group the terms into a trigonometric series in terms of *cos*(pA) and *sin*(pA) so that we have:

$$\frac{I}{\varepsilon_0 c_0} = \left|\vec{E}_0\right|^2(\alpha + \beta cos2A + \gamma sin2A + \eta cos4A + \cdots)$$

with:

$$\alpha = \frac{3+4cos2\Psi}{8};\ \beta = \frac{cos2\Psi}{2};\ \gamma = \frac{3sin2\Psi\cos(\Delta+\varphi)}{4};\ \eta = -\frac{3}{8}cos^2\Psi + sin^2\Psi$$

4) Explain how an FFT (Fast Fourier Transform) analysis of the I_A signal makes it possible to go back to the values of the ellipsometric parameters (Ψ, Δ).

Solution:

By locating the terms at 2 A, i.e. β and γ in the frequency space, we can determine the ellipsometric parameters Δ and Ψ. Beforehand, the device must be calibrated with a substrate in order to correctly adjust the parameters β and γ.

1.5. Appendices

1.5.1. *Conventions used for Jones vectors and Jones ABCD matrices*

The electric field, in the most general case, consists of an E_p component parallel to the plane of incidence and an E_s component perpendicular to the plane of incidence. The convention given in Figure 1.34 will be adopted such that ((Ox, Oy, Oz) ≡ (Os, Op, Oz)).

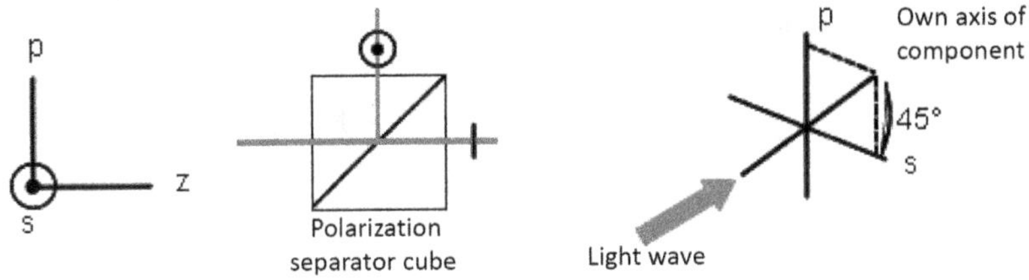

Figure 1.34. *Reference mark and sign convention to be adopted. For a color version of this figure, see www.iste.co.uk/dahoo/metrology2.zip*

Polarization *S* and *P*: $\vec{e}_S = \begin{pmatrix}1\\0\end{pmatrix}$ $\vec{e}_P = \begin{pmatrix}0\\1\end{pmatrix}$

An electromagnetic wave in the base defined by $\boldsymbol{e}_s$ and $\boldsymbol{e}_p$:

$$\vec{E}_L = \begin{pmatrix}|E_S|\\|E_P|\end{pmatrix} e^{-i\omega t}$$

A plane electromagnetic wave of zero phase polarized at 45° $\vec{E}_L = \vec{E}_S + \vec{E}_P$.

A base change matrix by angle rotation θ in a plane

$$R(\theta) = \begin{pmatrix}\cos\theta & -\sin\theta\\ \sin\theta & \cos\theta\end{pmatrix}$$

A reference change by rotation of θ $M(\theta) = R(\theta)M(0)R(-\theta)$.

A mirror at normal incidence $M = \begin{pmatrix}1 & 0\\0 & -1\end{pmatrix}$.

A separator 50/50 $LS = \begin{pmatrix}\frac{1}{2} & 0\\ 0 & \frac{1}{2}\end{pmatrix}$.

A polarizer cube oriented along *S* $CSP_S = \begin{pmatrix}1 & 0\\0 & 0\end{pmatrix}$.

A polarizer cube oriented according to *P* $CSP_P = \begin{pmatrix}0 & 0\\0 & 1\end{pmatrix}$.

Polarizers according to P and S: $P_0^P = \begin{pmatrix} 0 & 0 \\ 0 & 1 \end{pmatrix}$ $P_0^S = \begin{pmatrix} 1 & 0 \\ 0 & 0 \end{pmatrix}$.

A polarizer S oriented at -45: $P_{-45}^S = \frac{1}{2}\begin{pmatrix} 1 & -1 \\ -1 & 1 \end{pmatrix}$.

A quarter-wave plate along the proper axes:

$$Q_0 = e^{i\frac{\pi}{4}}\begin{pmatrix} 1 & 0 \\ 0 & -i \end{pmatrix} = e^{i\frac{\pi}{4}}\begin{pmatrix} 1 & 0 \\ 0 & e^{-i\frac{\pi}{2}} \end{pmatrix}$$

A half-wave plate along the proper axes: $L_0 = \begin{pmatrix} i & 0 \\ 0 & -i \end{pmatrix} = e^{i\frac{\pi}{2}}\begin{pmatrix} 1 & 0 \\ 0 & -1 \end{pmatrix}$.

A quarter wave plate oriented at 45° and -45°:

$$Q_{45} = \frac{1}{2}\begin{pmatrix} e^{-i\frac{\pi}{4}} & e^{i\frac{\pi}{4}} \\ e^{i\frac{\pi}{4}} & e^{-i\frac{\pi}{4}} \end{pmatrix} \quad Q_{-45} = \frac{1}{2}\begin{pmatrix} e^{-i\frac{\pi}{4}} & e^{-i\frac{3\pi}{4}} \\ e^{i\frac{3\pi}{4}} & e^{-i\frac{\pi}{4}} \end{pmatrix}$$

A phase shift matrix due to displacement δ for propagation in space:

$$D = \begin{pmatrix} \exp(i\phi) & 0 \\ 0 & \exp(i\phi) \end{pmatrix} \text{ with } \phi = nk_0\delta = 2\pi n\delta/\lambda_0$$

A phase shift matrix due to the Doppler effect along the p axis:

$$D_d = \begin{pmatrix} 0 & 0 \\ 0 & exp(-i\omega_d t) \end{pmatrix}$$

with $\omega_d = \omega_0\ (1 \pm 2nv/c)$, v being the speed of movement of the object.

A matrix representing the sample that changes the polarization of light in an ellipsometric setup: $SE = \begin{pmatrix} sin(\Psi)\, exp(i\nabla) & 0 \\ 0 & cos(\Psi) \end{pmatrix}$.

1.5.2. *2×2 transfer dies*

Propagation in homogeneous matter over a distance d:

$$M = \begin{bmatrix} 1 & d \\ 0 & 1 \end{bmatrix}$$

Propagation through a diopter separating two media with indices n_1 and n_2:n_1 and n_2:

$$M = \begin{bmatrix} 1 & 0 \\ 0 & \frac{n_1}{n_2} \end{bmatrix}$$

Transmission through a thin lens of focal length f ($f > 0$ convex, $f < 0$ concave):

$$M = \begin{bmatrix} 1 & 0 \\ -\frac{1}{f} & 1 \end{bmatrix}$$

Reflection on a concave mirror with a radius of curvature R:

$$M = \begin{bmatrix} 1 & 0 \\ \frac{2}{R} & 1 \end{bmatrix}$$

1.5.3. *2×2 matrix multiplication*

It is customary to use our matrix elements with the letters A, B, C and D:

$$M_2 = \begin{bmatrix} A_2 & C_2 \\ B_2 & D_2 \end{bmatrix} M_1 = \begin{bmatrix} A_1 & C_1 \\ B_1 & D_1 \end{bmatrix}$$

Identity matrix

$$I = \begin{bmatrix} 1 & 0 \\ 0 & 1 \end{bmatrix}$$

Multiplication of M_1 and M_2

$$M_1 M_2 = \begin{bmatrix} A_1 & C_1 \\ B_1 & D_1 \end{bmatrix} \begin{bmatrix} A_2 & C_2 \\ B_2 & D_2 \end{bmatrix} = \begin{bmatrix} A_1 A_2 + C_1 B_2 & A_1 C_2 + C_1 D_2 \\ B_1 A_2 + D_1 B_2 & B_1 C_2 + D_1 D_2 \end{bmatrix}$$

Determinant and inverse of the matrix M

$$\det(M) = \begin{vmatrix} A & C \\ B & D \end{vmatrix} = AD - BC \quad M^{-1} = \frac{1}{\det(M)} \begin{bmatrix} D & -C \\ -B & A \end{bmatrix}$$

1.5.4. *Trigonometric forms*

$$\sin^2(x) = (1 - cos2x)/2$$

$$\cos^2(x) = (1 + cos2x)/2$$

$$\sin^3(x) = \frac{sinx - sinxcos2x}{2} = (cosx - \frac{1}{2}sinx + \frac{sin3x}{2})/2$$

$$\cos^3(x) = \frac{cosx + cosxcos2x}{2} = (cosx + \frac{1}{2}cosx + \frac{cos3x}{2})/2$$

$$\sin^4(x) = (3 - 4cos2x + cos4x)/8$$

$$\cos^4(x) = (3 + 4cos2x + cos4x)/8$$

$$\cos(x)\cos(y) = \frac{\cos(x-y)-\cos(x+y)}{2} \quad \sin(x)\sin(y) = \frac{\cos(x-y)-\cos(x+y)}{2}$$

$$\sin(x)\cos(y) = \frac{\sin(x+y)+\sin(x-y)}{2} \quad \cos(x)\sin(y) = \frac{\sin(x+y)-\sin(x-y)}{2}$$

1.5.5. *Solution by MATLAB (exercises 1.4.3, 1.4.4 and 1.4.5)*

MATLAB

```
clear all
% Exercice 543
% Matrice de Jones
CSPS=[1 0; 0 0]
Q45=0.5*[1+1i 1-1i;1-1i 1+1i]
M2=[1 0; 0 -1]
Qm45=0.5*[1+1i -1+1i;-1+1i 1+1i]
CSPP=[0 0; 0 1]
PS45=0.5*[1 1;1 1]
%
%543
%
MS=PS45*CSPP*Qm45*M2*Q45*CSPS
MP=PS45*CSPS*Qm45*M2*Q45*CSPP
%
% expressions symbolic
% syms
% help syms
%
syms phirp  phirs
Dphirs=[exp(1i*phirs) 0;0 exp(1i*phirs)]
Dphirp=[exp(1i*phirp) 0;0 exp(1i*phirp)]
M843SPD=MS*Dphirs
M843PPD=MP*Dphirp
%
%
syms phis0 phip0 Es Ep
EL=[Es*exp(i*phis0) Ep*exp(i*phip0)]
ELt=transpose(EL)
%
%543
%
ESPD=M843SPD*ELt
EPPD=M843PPD*ELt
%
ER543=ESPD+EPPD
%
```

Solution

CSPS =

```
     1     0
     0     0
```

For a color version of this solution, see www.iste.co.uk/dahoo/metrology2.zip.

```
Q45 =
        0.5000 + 0.5000i   0.5000 - 0.5000i
        0.5000 - 0.5000i   0.5000 + 0.5000i
M2 =
        1    0
        0   -1
Qm45 =
        0.5000 + 0.5000i  -0.5000 + 0.5000i
       -0.5000 + 0.5000i   0.5000 + 0.5000i
CSPP =
        0    0
        0    1
PS45 =
        0.5000    0.5000
        0.5000    0.5000
MS =
       -0.5000         0
       -0.5000         0
MP =
        0    0.5000
        0    0.5000
Dphirs =
       [ exp(i*phirs),            0]
       [            0, exp(i*phirs)]

Dphirp =
       [ exp(i*phirp),            0]
       [            0, exp(i*phirp)]
M543SPD =
       [ -1/2*exp(i*phirs),                 0]
       [ -1/2*exp(i*phirs),                 0]
M543PPD =
       [                 0, 1/2*exp(i*phirp)]
       [                 0, 1/2*exp(i*phirp)]
EL =
       [ Es*exp(i*phis0), Ep*exp(i*phip0)]
ELt =
       [Es*exp(i*phis0) ]
       [Ep*exp(i*phip0) ]
```

ESPD =

[-1/2*exp(i*phirs)*Es*exp(i*phis0)]
[-1/2*exp(i*phirs)*Es*exp(i*phis0)]

EPPD =

[1/2*exp(i*phirp)*Ep*exp(i*phip0)]
[1/2*exp(i*phirp)*Ep*exp(i*phip0)]

ER843 =

[-1/2*exp(i*phirs)*Es*exp(i*phis0)+1/2*exp(i*phirp)*Ep*exp(i*phip0)]
[-1/2*exp(i*phirs)*Es*exp(i*phis0)+1/2*exp(i*phirp)*Ep*exp(i*phip0)]

MATLAB

```
%EXERCICE 544
clear all
% Matrice de Jones
CSPS=[1 0; 0 0]
Q45=0.5*[1+1i 1-1i;1-1i 1+1i]
M2=[1 0; 0 -1]
Qm45=0.5*[1+1i -1+1i;-1+1i 1+1i]
CSPP=[0 0; 0 1]
PS45=0.5*[1 1;1 1]
MS=PS45*CSPP*Qm45*M2*Q45*CSPS
MP=PS45*CSPS*Qm45*M2*Q45*CSPP
% expressions symbolic
syms phirp  phirs
Dphirs=[exp(1i*phirs) 0;0 exp(1i*phirs)]
Dphirp=[exp(1i*phirp) 0;0 exp(1i*phirp)]
MS544PD=MS*Dphirs
MP544PD=MP*Dphirp
%
%
syms phis phip Es Ep phi0  E0
%EL=[Es*exp(i*phis) Ep*exp(i*phip)]
EL=[E0*exp(i*phi0) E0*exp(i*phi0)]
ELt=transpose(EL)
ESPD=MS544PD*ELt
EPPD=MP544PD*ELt
%E0L=PS45*ELt
```

Solution

```
MS =
  -0.5000        0
  -0.5000        0
MP =
       0    0.5000
       0    0.5000
Dphirs =
           [ exp(i*phirs),           0]
           [           0, exp(i*phirs)]
Dphirp =
           [ exp(i*phirp),           0]
           [           0, exp(i*phirp)]
MS544PD =
           [ -1/2*exp(i*phirs),              0]
           [ -1/2*exp(i*phirs),              0]
MP544PD =
           [              0, 1/2*exp(i*phirp)]
           [              0, 1/2*exp(i*phirp)]
EL =
[ E0*exp(i*phi0), E0*exp(i*phi0)]
ELt =
           [ E0*exp(i*phi0) ]
           [ E0*exp(i*phi0) ]
ESPD =
           [ -1/2*exp(i*phirs)*E0*exp(i*phi0) ]
           [-1/2*exp(i*phirs)*E0*exp(i*phi0) ]
EPPD =
           [1/2*exp(i*phirp)*E0*exp(i*phi0) ]
           [1/2*exp(i*phirp)*E0*exp(i*phi0) ]
```

MATLAB

```
%
%EXO 545 Heterodyne
clear all
% Matrice de Jones
CSPS=[1 0; 0 0]
Q45=0.5*[1+1i 1-1i;1-1i 1+1i]
M2=[1 0; 0 -1]
Qm45=0.5*[1+1i -1+1i;-1+1i 1+1i]
CSPP=[0 0; 0 1]
PS45=0.5*[1 1;1 1]
MS=PS45*CSPS*CSPS
MP=PS45*CSPP*CSPP
```

```
% expressions symbolic
syms phirp  phirs
Dphirs=[exp(1i*phirs) 0;0 exp(1i*phirs)]
Dphirp=[exp(1i*phirp) 0;0 exp(1i*phirp)]
MS545=Dphirs*MS
MP545=Dphirp*MP
%
syms omegaD
DphiD=[0 0;0 exp(1i*omegaD)]
MP545=MP545*DphiD
%
syms omegas omegap Es Ep
EL=[Es*exp(i*omegas) Ep*exp(i*omegap)]
ELt=transpose(EL)
%
ESPD=MS545*ELt
EPPD=MP545*ELt
```

SOLUTION

```
MS545 =
        [ 1/2*exp(i*phirs),          0]
        [ 1/2*exp(i*phirs),          0]

MP545 =
        [          0, 1/2*exp(i*phirp)]
        [          0, 1/2*exp(i*phirp)]
DphiD =

        [        0,          0]
        [        0, exp(i*omegaD)]
MP545 =

        [                   0, 1/2*exp(i*phirp)*exp(i*omegaD)]
        [                   0, 1/2*exp(i*phirp)*exp(i*omegaD)]
EL =

        [ Es*exp(i*omegas), Ep*exp(i*omegap)]

ELt =
        [Es*exp(i*omegas) ]
        [Ep*exp(i*omegap) ]
```

```
ESPD =
        [1/2*exp(i*phirs)*Es*exp(i*omegas) ]
        [1/2*exp(i*phirs)*Es*exp(i*omegas) ]
EPPD =
        [1/2*exp(i*phirp)*exp(i*omegaD)*Ep*exp(i*omegap) ]
        [1/2*exp(i*phirp)*exp(i*omegaD)*Ep*exp(i*omegap) ]
```

1.6. Conclusion

In this chapter, a few examples have illustrated the application of Jones matrices to model the operation of interference devices in polarized light, in particular the homodyne, heterodyne interferometer or the ellipsometer. Some programs for MATLAB are given in the Appendix to simulate the operation of these devices.

2

Quantum-scale Interaction

To develop a model to interpret the properties of matter on the nanoscale, it is necessary to apply the principles of quantum mechanics. Thus, the effects of the close environment of a nanosystem are described by a statistical matrix, which introduces the coupling between the system and the surrounding environment. Two reference quantum models are used to study the effects of a heterogeneous medium on a nanosystem through the coupling at the volume and surface interfaces. They are based on a two-level quantum system interacting with an external field. This can be either a two-level laser system or a two-level magnetic spin system within the framework of the universal Ising-like model. These external fields can be an electromagnetic field, an electric field, a magnetic field, a thermal strain or a mechanical strain. The models applied to simulate the behavior of a nanosystem in the presence of the corresponding external field are often nonlinear and are generally solved by numerical methods.

2.1. Introduction

The study of the nature of phase transition and hence the corresponding changes of symmetry during phase change of first or second order covers a wide scientific field ranging from statistical physics to not only materials science but also computer science or socio-economical activities. In the field of condensed matter physics, the Ising model, the standard Potts model, the Blume–Capel model, the Bragg–Williams model, the Wajnflasz–Pick model or the Potts model, for instance, reveals the existence of possible invisible states through metastable states. There is no general method to bypass the so-called annealing method in systems that exhibit a phase transition. It is difficult to develop a method to monitor or avoid a type of phase transition in a target nanosystem.

The Ising model (1925) is easy to use and apply to other domains. The Ising model is used not only for the analysis of the phase transition observed in real materials but also in information science. In this case, the binary representation is the basic language. In fact, some difficulties in information science/engineering have been resolved from the point of view of statistical physics. An interdisciplinary science, which is the interface between statistical physics and information science, is developed in terms of the Ising model. Solving an optimization problem corresponds to finding the equilibrium state at finite temperature or the ground state of the Hamiltonian operator, which expresses the target optimization problem. In many cases, an optimization problem can be represented by the Ising model or its generalized model.

In statistical physics, the Ising model is used to study phase transitions in materials. The relationship between the universality class and the symmetry breaking at the transition point has been studied by developing models based on the Ising approach. Generally, the Hamiltonian operator H_0 of the Ising model, which includes a spin interaction over adjacent sites i and j and the interaction of the magnetic spin localized on the site (i) with an external magnetic field (h_i), is expressed by:

$$H_0 = -\sum_{\langle i,j \rangle}^{N} J_{ij}\sigma_{zi}\sigma_{zj} - \sum_{i=1}^{N} h_i \sigma_{zi}$$

where σ_{zi} is a microscopic spin variable (+1 or -1) at site i. Hereinafter, the g factor and the Bohr magneton (μ_B) are set to be one unit for simplicity (see exercise 2.4.1).

If $Jij > 0$, the interaction between the ith and jth sites is ferromagnetic.

If $Jij < 0$, the interaction is antiferromagnetic.

Generally, the energy hypersurface of optimization problems is complex as random spin systems problems. The simulated annealing method has been proposed by Kirkpatrick using an algorithm, which in most cases solves the optimization problem. In simulated annealing, the temperature is first set to a maximum value. Then, the temperature is gradually lowered to the best or most plausible solution. Given a current state i of the solid with energy E_i and a subsequent state j of energy E_j, a characteristic transition time (τ) at temperature T is expressed as $\tau \propto \beta E$, where β denotes the inverse temperature ($\beta = T^{-1}$) and E represents the energy difference. $E = (E_j - E_j)$. For the sake of simplicity, the Boltzmann constant k_B is set to the unit value.

In the case of a two-level laser system interacting with a medium, the relaxation and decoherence processes that are considered result from noise sources or inhomogeneous broadening in an ensemble of atoms. This approach is illustrated by studying the behavior of a two-level system illuminated by coherent light.

2.2. The spin through the Dirac equation

2.2.1. *Theoretical background*

In the theory of electromagnetism, the wave equation is invariant in a Lorentz transformation and leads to the dispersion equation: $\frac{\omega^2}{c^2} = k^2$

In a non-relativistic approach, the kinetic energy of a free particle is given in terms of its momentum (p) and its mass (m) by:

$$E = \frac{p^2}{2m}$$

Using the correspondence between differential operators and the energy operator $\left(E \rightarrow i\hbar\frac{\partial}{\partial t}\right)$ and the momentum operator $\left(p \rightarrow \frac{\hbar}{i}\nabla\right)$, the time-dependent Schrödinger equation corresponding to $E = \frac{p^2}{2m}$ is expressed by:

$$i\hbar\frac{\partial}{\partial t}\Psi(r,t) = -\frac{\hbar}{2m}\Delta\Psi(r,t)$$

Taking a wave function in the form of a plane wave such as: $\Psi(r,t) = e^{i(pr-Et)/\hbar}$, which is a solution of the Schrödinger equation, the classical relation $E = \frac{p^2}{2m}$ is obtained. However, the Schrödinger equation, which includes first-order time derivatives and second-order space derivatives, is not Lorentz invariant.

However, a Lorentz invariant equation can be obtained if the previous operators are applied to the relativistic energy equation:

$$E^2 = p^2c^2 + m^2c^4$$

This leads to the Klein–Gordon equation:

$$-\hbar^2 \frac{\partial^2}{\partial t^2}\Psi(r,t) = -\hbar^2 c^2 \Delta\Psi(r,t) + m^2c^4\Psi(r,t)$$

An important difference between the Klein–Gordon equation and the Schrödinger equation is that the time derivative is of the second order.

In quantum mechanics, although a plane wave is a solution of the Schrödinger equation, it is not a quantum state as it is not normalized and does not belong to a Hilbert space. However, a free particle of energy E and momentum p can be represented by a wave function $\Psi(r,t)$ consisting of superposed monochromatic plane waves of frequency $E/\hbar$ and wave vector $p/\hbar$ such as:

$$\Psi(r,t) = \frac{1}{(2\pi\hbar)^{\frac{3}{2}}} \int F(p) e^{i(pr-Et)/\hbar} dp$$

In the case of the Schrödinger equation, the probability density ρ is given by:

$\rho = \Psi\Psi^*$, and the total probability is normalized such that $\int \rho d^3r = 1$.

Because $\frac{\partial \Psi}{\partial t}$, Ψ and their conjugates $\frac{\partial \Psi^*}{\partial t}, \Psi^*$ are dependent through the Schrödinger equation:

$$\frac{\partial}{\partial t}\int \rho d^3 r = \frac{\partial}{\partial t}\int \Psi\Psi^* d^3 r = 0$$

This means that the total probability of the particle is independent of time.

The Schrödinger equation can be transformed into the equation of continuity $\vec{\nabla}.\vec{j} + \frac{\partial \rho}{\partial t} = 0$ giving the probability density as $\rho = \Psi\Psi^*$ and the probability current density as $\vec{j} = \frac{i\hbar}{2m}\left(\Psi\vec{\nabla}\Psi^* - \Psi^*\vec{\nabla}\Psi\right)$.

In the case of the Klein–Gordon equation, because $\frac{\partial \Psi}{\partial t}$ and Ψ and their conjugates are independent:

$$\frac{\partial}{\partial t}\int \Psi\Psi^* d^3 r = \int\left(\frac{\partial \Psi}{\partial t}\Psi^* + \frac{\partial \Psi^*}{\partial t}\Psi\right) d^3 r \neq 0$$

In this case, the expression of the probability density cannot be expressed as $\Psi\Psi^*$.

For a wave function $\Psi(r,t)$, the Klein–Gordon equation in free space when no field is applied is written as:

$$-E^2\Psi = -\left|\vec{p}^2\right|c^2\Psi - m^2c^4\Psi$$

The solutions are given by: $E = \pm\sqrt{\sqrt{\left|\vec{p}^2\right|} + m^2c^4}$.

Consequently, solutions corresponding to particles of negative energy are possible.

The Klein–Gordon equation can also be used to calculate the number ρ of particles per unit volume and the current density $\vec{j}$ representing the flow of particles passing through a unit area per unit time, which obeys the continuity equation:

$$\vec{\nabla}.\vec{j}+\frac{\partial\rho}{\partial t}=0$$

In this case, the probability density is written as: $\rho=\frac{i\hbar}{2mc^2}\left(\Psi\frac{\partial\Psi^*}{\partial t}-\Psi^*\frac{\partial\Psi}{\partial t}\right)$ and the probability current density is as before. But the expression of the probability density is not positive as a probability density should be.

For a plane wave $\Psi(r,t)=Ne^{i(\vec{p}.\vec{r}-Et)}$, the number of particles per unit volume is written as:

$$\rho=\frac{i\hbar}{2mc^2}\left(\Psi^*\frac{\partial\Psi}{\partial t}-\Psi\frac{\partial\Psi^*}{\partial t}\right)=\frac{i\hbar}{2mc^2}\left(-\frac{i}{\hbar}EN^2-\frac{i}{\hbar}EN^2\right)=\frac{\left(2N^2E\right)}{2mc^2}$$

and the current density is written as:

$$\vec{j}=\frac{\hbar}{2mi}\left(\Psi^*\vec{\nabla}\Psi-\Psi\vec{\nabla}\Psi^*\right)=\frac{\hbar}{2mi}\left(\frac{i\vec{p}}{\hbar}N^2+\frac{i\vec{p}}{\hbar}N^2\right)=N^2\frac{\vec{p}}{m}$$

For particles of negative energy, the probability densities of particles ρ are negative which is not physically meaningful as pointed out above. In fact, the Klein–Gordon equation is equivalent to two simultaneous equations operating in a 2D space where each wave function is a two-component spinor such as:

$$\Psi(r,t)=\begin{pmatrix}\phi(r,t)\\ \chi(r,t)\end{pmatrix}$$

and

$$i\hbar \frac{\partial}{\partial t}\phi(r,t) = -\frac{\hbar^2}{2m}\Delta\big(\phi(r,t)+\chi(r,t)\big)+m\,c^2\phi(r,t)$$

$$i\hbar \frac{\partial}{\partial t}\chi(r,t) = +\frac{\hbar^2}{2m}\Delta\big(\phi(r,t)+\chi(r,t)\big)-m\,c^2\chi(r,t)$$

The probability density is then given by:

$$\rho = |\phi|^2 - |\chi|^2$$

This density is still not positive definite. But if it is interpreted as a charge density instead of a particle density, then it is positive definite. In this case, ρ and $\vec{j}$ are multiplied by e (the unit of charge). $|\phi(r)|^2$ and $|\chi(r)|^2$ are interpreted as the probability of finding the particle at r in the positive and negative charge spaces, respectively. The special spinors $\Psi_+(r,t)=\begin{pmatrix}\phi(r,t)\\ 0\end{pmatrix}$ and $\Psi_-(r,t)=\begin{pmatrix}0\\ \chi(r,t)\end{pmatrix}$ represent purely positive and negative charge state, respectively. For a given charge particle with wave function ϕ and a positive charge density, it is necessary to postulate the existence of a corresponding antiparticle wave function χ and a negative charge density. This is obtained straightforward with the Dirac equation.

PAM Dirac looked for solutions to the relativistic equation of the energy of a free particle (the electron) [DIR 28] by using an equation comprising only time derivatives of the first order and of which the Hamiltonian of the free particle linearly depends on the momentum and the mass.

Representing the solution wave function by the ket $|\Psi\rangle$, Dirac poses the equation: $i\frac{\partial}{\partial t}|\Psi\rangle = \hat{H}|\Psi\rangle = (\bar{\alpha}.\bar{p}+\beta mc)|\Psi\rangle$ and then establishes the conditions that the vector $\bar{\alpha}=(\alpha_x,\alpha_y,\alpha_z)$ and the scalar β must satisfy $\bar{\alpha}.\bar{p}+\beta mc$ in order for it to be a solution. For this solution to be acceptable, the free particle must also satisfy the relativistic energy equation $E^2 = p^2c^2 + m^2c^4$, i.e. the Klein–Gordon equation.

This theoretical approach results in particles whose energy can be negative and provides particle densities that are all positive. Applied to the case of the electron, the solutions of the Dirac equation also describe the spin and the magnetic moment.

2.2.2. *Application: the Dirac equation and Pauli matrices*

1) If $\frac{E}{c}=\sqrt{p^2+m^2c^2}=(\bar{\alpha}.\bar{p}+\beta mc)$, then $(\alpha p+\beta mc)(\alpha p+\beta mc)=$ $p^2+m^2c^4$.

Setting $\bar{p}=(p_x,p_y,p_z)$ and $\bar{\alpha}=(\alpha_x,\alpha_y,\alpha_z)$ establishes the conditions that must be fulfilled by the square of the components of the vector and of the vector α and β , the cross terms of the vectors α and β for the above equation to be satisfied.

2) Demonstrate that the conditions obtained in question 1 impose that α_i and β are matrices and not scalars.

3) α_i and β must be matrices. These matrices act on an abstract space of at least 4 dimensions. Dealing with 4 x 4 matrices. Dirac used the Pauli matrices to construct the corresponding 4 x 4 matrices and called this four-dimensional space the space of spinors.

If *I* denotes the 2 x 2 identity matrix, 0 the null 2 x 2 matrix and σ_i the Hermitian–Pauli matrices ([DAH 16] page 93), demonstrate that the matrices α_i and β are hermitic and verify the following relations, i.e.:

$$\alpha_x^2,=\alpha_y^2=\alpha_z^2=\beta^2=1$$

$$\alpha_i\beta+\beta\alpha_i=0$$

$$\alpha_j\alpha_k+\alpha_k\alpha_j=0 \quad j\neq k$$

Reminder: a Hermitian matrix is a square matrix with complex elements that satisfies the following property: the matrix is equal to its conjugate transposed matrix. In other words, $a_{ij} = a^*_{ji}$

4) The form of the Dirac equation can be simplified by multiplying the two terms of the equation by β and using the γ^i Dirac matrices which for $i = 0,1,2,3$ are defined by:

$$\gamma^0 = \beta, \qquad \gamma^2 = \beta\alpha_y \quad \gamma^3 = \beta\alpha_z$$

Express the γ^i Dirac matrices as a 4 x 4 matrix and as 2 x 2 matrices using the σ_i Pauli matrices

$$\sigma_x = \begin{pmatrix} 0 & 1 \\ 1 & 0 \end{pmatrix} \quad \sigma_y = \begin{pmatrix} 0 & -i \\ i & 0 \end{pmatrix} \quad \sigma_z = \begin{pmatrix} 1 & 0 \\ 0 & -1 \end{pmatrix}$$

5) Paul Dirac sought to solve the relativistic Schrödinger equation describing the wave function of a fermion of mass m, energy E and momentum p in the absence of an electromagnetic field:

$$i\hbar\frac{\partial}{\partial t}|\Psi\rangle = H|\Psi\rangle = (\alpha.p + \beta m)|\Psi\rangle$$

By setting $c = 1$ and $\hbar = 1$, multiply by β the two sides of this equation and obtain a simplified relativistic equation using the Dirac matrices, γ^0, γ^1, γ^2 and γ^3.

Express the obtained system of equations.

6) The solutions can be written in the form of plane waves $|\Psi\rangle = u(E,\vec{p})e^{i(\vec{p}.\vec{r}-Et)}$, where $u(E,\vec{p})$ is a four-component spinor. Find the solutions of the Dirac equation for a particle at rest for which $\vec{p} = 0$.

7) Find the solutions of a free particle in motion in the form of plane waves $|\Psi\rangle = u(E,\vec{p})e^{i(\vec{p}.\vec{r}-Et)}$ by writing $u(E,\vec{p}) = \begin{pmatrix} u_a \\ u_b \end{pmatrix}$.

8) Verify that the kets $|\Psi^1\rangle$, $|\Psi^2\rangle$, $|\Psi^3\rangle$ and $|\Psi^4\rangle$ (defined below) are solutions of this system of equations:

$$|\Psi^1\rangle = u_1 e^{-ipr} = \sqrt{\frac{2E}{E+m}} \begin{pmatrix} 1 \\ 0 \\ \dfrac{p^3}{E+m} \\ \dfrac{p^1+ip^2}{E+m} \end{pmatrix} e^{-ipr}$$

$$|\Psi^2\rangle = u_2 e^{-ipr} = \sqrt{\frac{2E}{E+m}} \begin{pmatrix} 0 \\ 1 \\ \dfrac{p^1-ip^2}{E+m} \\ \dfrac{-p^3}{E+m} \end{pmatrix} e^{-ipr}$$

$$|\Psi^3\rangle = u_3 e^{ipr} = \sqrt{\frac{2E}{E-m}} \begin{pmatrix} \dfrac{p^3}{E-m} \\ \dfrac{p^1+ip^2}{E-m} \\ 1 \\ 0 \end{pmatrix} e^{ipr}$$

$$|\Psi^4\rangle = u_4 e^{ipr} \sqrt{\frac{2E}{E-m}} \begin{pmatrix} \dfrac{p^1-ip^2}{E-m} \\ \dfrac{-p^3}{E-m} \\ 0 \\ 1 \end{pmatrix} e^{ipr}$$

9) According to Stückelberg–Feynman interpretation, a negative energy state corresponds to an antiparticle state with a positive energy E moving backwards in time so having a reverse sign p.

Based on this interpretation, derive the expressions of the kets $|\Psi^3\rangle = v_2(p)$ and $|\Psi^4\rangle = v_1(p)$ corresponding to positive-energy antiparticles.

10) Check that with the chosen normalization coefficient:

– the dot product of u_1 with itself is equal to 1;

– the scalar product of u_2 with itself is equal to 1;

– the scalar product of v_2 with itself is equal to 1;

– the scalar product of v_1 with itself is equal to 1.

11) Check that u_1 and u_2, v_1 and v_2, $u_1(p)$ and $v_2(-p)$, $u_2(p)$ and $v_1(-p)$ are orthogonal:

2.2.2.1. *Answer to question 1*

As: $(\alpha p + \beta mc)(\alpha p + \beta mc) = p^2 + m^2c^4$

$$\left(\alpha_x p_x + \alpha_y p_y + \alpha_z p_z + \beta mc\right)\left(\alpha_x p_x + \alpha_y p_y + \alpha_z p_z + \beta mc\right) = p^2 + m^2c^4$$

As each component of the momentum operator verifies the following commutation relations:

$$p_x p_z = p_z p_x$$

$$p_y p_z = p_z p_y$$

$$p_x p_y = p_y p_x$$

The developed equation can be put in the following form by grouping the commuting terms:

$$\Rightarrow \alpha_x^2 p_x^2 + \alpha_y^2 p_y^2 + \alpha_z^2 p_z^2 + \beta^2 m^2 c^4 +$$
$$\left(\alpha_x\alpha_y + \alpha_y\alpha_x\right) p_x p_y + \left(\alpha_x\alpha_z + \alpha_z\alpha_x\right) p_x p_z + \left(\alpha_y\alpha_z + \alpha_z\alpha_y\right) p_y p_z +$$
$$\left(2\alpha_x\beta\right) p_x mc^2 + \left(2\alpha_y\beta\right) p_y mc^2 + \left(2\alpha_z\beta\right) p_z mc^2 = p^2 + m^2c^4$$

To satisfy equality, the terms $\alpha_x^2, \alpha_y^2, \alpha_z^2$ and β^2 must be equal to 1. That is to say: $\alpha_i^2 = 1, \quad i = 1,2,3$ and $\beta = 1$.

and:

$$\left(\alpha_x\alpha_y + \alpha_y\alpha_x\right) p_x p_y + \left(\alpha_x\alpha_z + \alpha_z\alpha_x\right) p_x p_z + \left(\alpha_y\alpha_z + \alpha_z\alpha_y\right) p_y p_z = 0$$

and $\left(2\alpha_x\beta\right) p_x mc^2 + \left(2\alpha_y\beta\right) p_y mc^2 + \left(2\alpha_z\beta\right) p_z mc^2 = 0$

That is to say, the following anti-commutation relations:

$$\left(\alpha_x\alpha_y + \alpha_y\alpha_x\right) = 0, \left(\alpha_x\alpha_z + \alpha_z\alpha_x\right) = 0, \left(\alpha_y\alpha_z + \alpha_z\alpha_y\right) = 0$$

and the following multiplication results:

$$\alpha_x\beta = 0, \quad \alpha_y\beta = 0, \quad \alpha_z\beta = 0$$

2.2.2.2. *Answer to question 2*

Equations $\left(\alpha_i\alpha_j + \alpha_j\alpha_i\right) = 0, \; i \neq j$ mean $\alpha_x, \; \alpha_y, \; \alpha_z, \beta$ anti-commute.

If $\alpha_x, \; \alpha_y, \; \alpha_z$ were scalars (numbers), they would necessarily commute. This would imply that all α_i are zero.

The only possibility is that α_i are matrices of operators acting on the ket $|\Psi\rangle$.

If β were a scalar (a number), the equations $\alpha_x\beta=0$, $\alpha_y\beta=0$, $\alpha_z\beta=0$ would imply that β is zero. The only possibility is that β it is also a matrix of an operator acting on the ket $|\Psi\rangle$.

2.2.2.3. *Answer to question 3*

From the form given for the matrices α_i and β which have for expression $\beta=\begin{pmatrix} I & 0 \\ 0 & -I \end{pmatrix}$ and $\alpha_i=\begin{pmatrix} 0 & \sigma_i \\ \sigma_i & 0 \end{pmatrix}$, the matrices α_i are written as:

$$\alpha_x=\begin{pmatrix} 0 & 0 & 0 & 1 \\ 0 & 0 & 1 & 0 \\ 0 & 1 & 0 & 0 \\ 1 & 0 & 0 & 0 \end{pmatrix}, \alpha_y=\begin{pmatrix} 0 & 0 & 0 & -i \\ 0 & 0 & i & 0 \\ 0 & -i & 0 & 0 \\ i & 0 & 0 & 0 \end{pmatrix}, \alpha_z=\begin{pmatrix} 0 & 0 & 1 & 0 \\ 0 & 0 & 0 & -1 \\ 1 & 0 & 0 & 0 \\ 0 & -1 & 0 & 0 \end{pmatrix}$$

The matrix β is written as: $\beta=\begin{pmatrix} 1 & 0 & 0 & 0 \\ 0 & 1 & 0 & 0 \\ 0 & 0 & -1 & 0 \\ 0 & 0 & 0 & -1 \end{pmatrix}$

These four matrices have a zero trace (sum of the diagonal elements). The main diagonal of the matrices α_i consists of zeros. The matrix β is a square matrix with zero non-diagonal elements and real diagonal elements. It is equal to its conjugate transpose and therefore it is Hermitian. The calculations prove that:

$$\beta^2=\begin{pmatrix} 1 & 0 & 0 & 0 \\ 0 & 1 & 0 & 0 \\ 0 & 0 & -1 & 0 \\ 0 & 0 & 0 & -1 \end{pmatrix}\cdot\begin{pmatrix} 1 & 0 & 0 & 0 \\ 0 & 1 & 0 & 0 \\ 0 & 0 & -1 & 0 \\ 0 & 0 & 0 & -1 \end{pmatrix}=\begin{pmatrix} 1 & 0 & 0 & 0 \\ 0 & 1 & 0 & 0 \\ 0 & 0 & 1 & 0 \\ 0 & 0 & 0 & 1 \end{pmatrix}=1$$

In the case of matrices α_i, the elements a_{ij} and a_{ji} verify that $a_{ij} = a_{ji}^*$. These matrices are hermitic. The $\alpha_x^2, \alpha_y^2, \alpha_z^2$ matrices are equal to:

$$\alpha_x^2 = \begin{pmatrix} 0 & 0 & 0 & 1 \\ 0 & 0 & 1 & 0 \\ 0 & 1 & 0 & 0 \\ 1 & 0 & 0 & 0 \end{pmatrix} \cdot \begin{pmatrix} 0 & 0 & 0 & 1 \\ 0 & 0 & 1 & 0 \\ 0 & 1 & 0 & 0 \\ 1 & 0 & 0 & 0 \end{pmatrix} = \begin{pmatrix} 1 & 0 & 0 & 0 \\ 0 & 1 & 0 & 0 \\ 0 & 0 & 1 & 0 \\ 0 & 0 & 0 & 1 \end{pmatrix} = 1$$

$$\alpha_y^2 = \begin{pmatrix} 0 & 0 & 0 & -i \\ 0 & 0 & i & 0 \\ 0 & -i & 0 & 0 \\ i & 0 & 0 & 0 \end{pmatrix} \cdot \begin{pmatrix} 0 & 0 & 0 & -i \\ 0 & 0 & i & 0 \\ 0 & -i & 0 & 0 \\ i & 0 & 0 & 0 \end{pmatrix} = \begin{pmatrix} 1 & 0 & 0 & 0 \\ 0 & 1 & 0 & 0 \\ 0 & 0 & 1 & 0 \\ 0 & 0 & 0 & 1 \end{pmatrix} = 1$$

$$\alpha_z^2 = \begin{pmatrix} 0 & 0 & 1 & 0 \\ 0 & 0 & 0 & -1 \\ 1 & 0 & 0 & 0 \\ 0 & -1 & 0 & 0 \end{pmatrix} \cdot \begin{pmatrix} 0 & 0 & 1 & 0 \\ 0 & 0 & 0 & -1 \\ 1 & 0 & 0 & 0 \\ 0 & -1 & 0 & 0 \end{pmatrix} = \begin{pmatrix} 1 & 0 & 0 & 0 \\ 0 & 1 & 0 & 0 \\ 0 & 0 & 1 & 0 \\ 0 & 0 & 0 & 1 \end{pmatrix} = 1$$

Similarly, it can be shown that:

$$\alpha_x \beta = \begin{pmatrix} 0 & 0 & 0 & 1 \\ 0 & 0 & 1 & 0 \\ 0 & 1 & 0 & 0 \\ 1 & 0 & 0 & 0 \end{pmatrix} \cdot \begin{pmatrix} 1 & 0 & 0 & 0 \\ 0 & 1 & 0 & 0 \\ 0 & 0 & -1 & 0 \\ 0 & 0 & 0 & -1 \end{pmatrix} = \begin{pmatrix} 0 & 0 & 0 & -1 \\ 0 & 0 & -1 & 0 \\ 0 & 1 & 0 & 0 \\ 1 & 0 & 0 & 0 \end{pmatrix}$$

$$\beta \alpha_x = \begin{pmatrix} 1 & 0 & 0 & 0 \\ 0 & 1 & 0 & 0 \\ 0 & 0 & -1 & 0 \\ 0 & 0 & 0 & -1 \end{pmatrix} \cdot \begin{pmatrix} 0 & 0 & 0 & 1 \\ 0 & 0 & 1 & 0 \\ 0 & 1 & 0 & 0 \\ 1 & 0 & 0 & 0 \end{pmatrix} = \begin{pmatrix} 0 & 0 & 0 & 1 \\ 0 & 0 & 1 & 0 \\ 0 & -1 & 0 & 0 \\ -1 & 0 & 0 & 0 \end{pmatrix}$$

$$\alpha_x\beta+\beta\alpha_x=\begin{pmatrix}0&0&0&-1\\0&0&-1&0\\0&1&0&0\\1&0&0&0\end{pmatrix}+\begin{pmatrix}0&0&0&1\\0&0&1&0\\0&-1&0&0\\-1&0&0&0\end{pmatrix}=0$$

$$\alpha_y\beta=\begin{pmatrix}0&0&0&-i\\0&0&i&0\\0&-i&0&0\\i&0&0&0\end{pmatrix}\cdot\begin{pmatrix}1&0&0&0\\0&1&0&0\\0&0&-1&0\\0&0&0&-1\end{pmatrix}=\begin{pmatrix}0&0&0&i\\0&0&-i&0\\0&-i&0&0\\i&0&0&0\end{pmatrix}$$

$$\beta\alpha_y=\begin{pmatrix}1&0&0&0\\0&1&0&0\\0&0&-1&0\\0&0&0&-1\end{pmatrix}\cdot\begin{pmatrix}0&0&0&-i\\0&0&i&0\\0&-i&0&0\\i&0&0&0\end{pmatrix}=\begin{pmatrix}0&0&0&-i\\0&0&i&0\\0&i&0&0\\-i&0&0&0\end{pmatrix}$$

$$\alpha_y\beta+\beta\alpha_y=\begin{pmatrix}0&0&0&i\\0&0&-i&0\\0&-i&0&0\\i&0&0&0\end{pmatrix}+\begin{pmatrix}0&0&0&-i\\0&0&i&0\\0&i&0&0\\-i&0&0&0\end{pmatrix}=0$$

$$\alpha_z\beta=\begin{pmatrix}0&0&1&0\\0&0&0&-1\\1&0&0&0\\0&-1&0&0\end{pmatrix}\cdot\begin{pmatrix}1&0&0&0\\0&1&0&0\\0&0&-1&0\\0&0&0&-1\end{pmatrix}=\begin{pmatrix}0&0&-1&0\\0&0&0&1\\1&0&0&0\\0&-1&0&0\end{pmatrix}0$$

$$\beta\alpha_z=\begin{pmatrix}1&0&0&0\\0&1&0&0\\0&0&-1&0\\0&0&0&-1\end{pmatrix}\cdot\begin{pmatrix}0&0&1&0\\0&0&0&-1\\1&0&0&0\\0&-1&0&0\end{pmatrix}=\begin{pmatrix}0&0&1&0\\0&0&0&-1\\-1&0&0&0\\0&1&0&0\end{pmatrix}$$

$$\alpha_z\beta+\beta\alpha_z=\begin{pmatrix}0&0&-1&0\\0&0&0&1\\1&0&0&0\\0&-1&0&0\end{pmatrix}+\begin{pmatrix}0&0&1&0\\0&0&0&-1\\-1&0&0&0\\0&1&0&0\end{pmatrix}=0$$

Finally:

$$\alpha_x\alpha_y+\alpha_y\alpha_x=\begin{pmatrix}0&0&0&1\\0&0&1&0\\0&1&0&0\\1&0&0&0\end{pmatrix}\cdot\begin{pmatrix}0&0&0&-i\\0&0&i&0\\0&-i&0&0\\i&0&0&0\end{pmatrix}+\begin{pmatrix}0&0&0&-i\\0&0&i&0\\0&-i&0&0\\i&0&0&0\end{pmatrix}\cdot\begin{pmatrix}0&0&0&1\\0&0&1&0\\0&1&0&0\\1&0&0&0\end{pmatrix}$$

$$\alpha_x\alpha_y+\alpha_y\alpha_x=\begin{pmatrix}i&0&0&0\\0&-i&0&0\\0&0&i&0\\0&0&0&-i\end{pmatrix}+\begin{pmatrix}-i&0&0&0\\0&i&0&0\\0&0&-i&0\\0&0&0&i\end{pmatrix}=0\;0$$

$$\alpha_x\alpha_z+\alpha_z\alpha_x=\begin{pmatrix}0&0&0&1\\0&0&1&0\\0&1&0&0\\1&0&0&0\end{pmatrix}\cdot\begin{pmatrix}0&0&1&0\\0&0&0&-1\\1&0&0&0\\0&-1&0&0\end{pmatrix}+\begin{pmatrix}0&0&1&0\\0&0&0&-1\\1&0&0&0\\0&-1&0&0\end{pmatrix}\cdot\begin{pmatrix}0&0&0&1\\0&0&1&0\\0&1&0&0\\1&0&0&0\end{pmatrix}$$

$$\alpha_x\alpha_z+\alpha_z\alpha_x=\begin{pmatrix}0&-1&0&0\\1&0&0&0\\0&0&0&-1\\0&0&1&0\end{pmatrix}+\begin{pmatrix}0&1&0&0\\-1&0&0&0\\0&0&0&1\\0&0&-1&0\end{pmatrix}=0$$

$$\alpha_y\alpha_z+\alpha_z\alpha_y=\begin{pmatrix}0&0&0&1\\0&0&1&0\\0&1&0&0\\1&0&0&0\end{pmatrix}\cdot\begin{pmatrix}0&0&1&0\\0&0&0&-1\\1&0&0&0\\0&-1&0&0\end{pmatrix}+\begin{pmatrix}0&0&1&0\\0&0&0&-1\\1&0&0&0\\0&-1&0&0\end{pmatrix}\cdot\begin{pmatrix}0&0&0&1\\0&0&1&0\\0&1&0&0\\1&0&0&0\end{pmatrix}$$

$$\alpha_y\alpha_z+\alpha_z\alpha_y=\begin{pmatrix}0&-1&0&0\\1&0&0&0\\0&0&0&-1\\0&0&1&0\end{pmatrix}+\begin{pmatrix}0&1&0&0\\-1&0&0&0\\0&0&0&1\\0&0&-1&0\end{pmatrix}=0$$

Therefore, the equations $\alpha_j\alpha_k+\alpha_k\alpha_j=0 \quad j\neq k$ are verified.

2.2.2.4. *Answer to question 4*

The matrix γ^0 is written as:

$$\gamma^0=\begin{pmatrix}1&0&0&0\\0&1&0&0\\0&0&-1&0\\0&0&0&-1\end{pmatrix}$$

$$\gamma^1=\begin{pmatrix}1&0&0&0\\0&1&0&0\\0&0&-1&0\\0&0&0&-1\end{pmatrix}\cdot\begin{pmatrix}0&0&0&1\\0&0&1&0\\0&1&0&0\\1&0&0&0\end{pmatrix}=\begin{pmatrix}0&0&0&1\\0&0&1&0\\0&-1&0&0\\-1&0&0&0\end{pmatrix}$$

$$\gamma^2=\begin{pmatrix}1&0&0&0\\0&1&0&0\\0&0&-1&0\\0&0&0&-1\end{pmatrix}\cdot\begin{pmatrix}0&0&0&-i\\0&0&i&0\\0&-i&0&0\\i&0&0&0\end{pmatrix}=\begin{pmatrix}0&0&0&-i\\0&0&i&0\\0&i&0&0\\-i&0&0&0\end{pmatrix}$$

$$\gamma^3=\begin{pmatrix}1&0&0&0\\0&1&0&0\\0&0&-1&0\\0&0&0&-1\end{pmatrix}\cdot\begin{pmatrix}0&0&1&0\\0&0&0&-1\\1&0&0&0\\0&-1&0&0\end{pmatrix}=\begin{pmatrix}0&0&1&0\\0&0&0&-1\\-1&0&0&0\\0&1&0&0\end{pmatrix}$$

Using the Pauli matrices and the identity matrix I, the matrices $\gamma^0, \gamma^1, \gamma^2, \gamma^3$ can be written as:

$$\gamma^0 = \begin{pmatrix} I & 0 \\ 0 & -I \end{pmatrix}$$

$$\gamma^2 = \begin{pmatrix} 0 & \sigma_y \\ -\sigma_y & 0 \end{pmatrix}$$

$$\gamma^3 = \begin{pmatrix} 0 & \sigma_z \\ -\sigma_z & 0 \end{pmatrix}$$

2.2.2.5. *Answer to question 5*

Setting $c = 1$ (speed of light) and $\hbar = 1$ (with h denoting the reduced Planck constant) by multiplying by β, the Dirac equation is obtained. This equation describes the wave function of a fermion of mass m, energy E and momentum p in the absence of an electromagnetic field:

$$i\beta\frac{\partial}{\partial t}|\Psi\rangle = \left(\beta\alpha_i p_i + \beta^2 m\right)|\Psi\rangle$$

$$i\gamma^0\frac{\partial}{\partial t}|\Psi\rangle = \left(\gamma^i p_i + m\right)|\Psi\rangle$$

where i = 1.2.3 and where Einstein's convention is used (the indices that are repeated are summed).

With the convention $\frac{\partial}{\partial \mu} = \left(\frac{\partial}{\partial t}, \frac{\partial}{\partial x}, \frac{\partial}{\partial y}, \frac{\partial}{\partial z}\right)$, where t represents time and x, y and z represent the coordinates in space, the Dirac equation is written as:

$$i\gamma^0\frac{\partial}{\partial t}|\Psi\rangle = \left(\gamma^i p_i + m\right)|\Psi\rangle = \left(-i\gamma^i\frac{\partial}{\partial x^i} + m\right)|\Psi\rangle$$

$$i\gamma^0 \frac{\partial}{\partial_0}|\Psi\rangle = \left(-i\gamma^i \frac{\partial}{\partial_i} + m\right)|\Psi\rangle \Rightarrow i\gamma^\nu \partial_\nu |\Psi_\eta\rangle = m|\Psi_\eta\rangle$$

$$\eta = 1, 2, 3, 4 \quad and \quad \nu = 0, 1, 2, 3$$

$$i\gamma^0 \partial_0 |\Psi_\eta\rangle = i\partial_0 \begin{pmatrix} 1 & 0 & 0 & 0 \\ 0 & 1 & 0 & 0 \\ 0 & 0 & -1 & 0 \\ 0 & 0 & 0 & -1 \end{pmatrix} |\Psi_\eta\rangle$$

If $a_{n1}, a_{n2}, a_{n3}, a_{n4}$ represent the components of ket $|\Psi_n\rangle$ in the space of spinors, then:

$$i\gamma^0 \partial_0 |\Psi_1\rangle = i\partial_0 a_{11}$$
$$i\gamma^0 \partial_0 |\Psi_2\rangle = i\partial_0 a_{22}$$
$$i\gamma^0 \partial_0 |\Psi_3\rangle = -i\partial_0 a_{33}$$
$$i\gamma^0 \partial_0 |\Psi_4\rangle = -i\partial_0 a_{44}$$

$$i\gamma^1 \partial_1 |\Psi_\eta\rangle = i\partial_1 \begin{pmatrix} 0 & 0 & 0 & 1 \\ 0 & 0 & 1 & 0 \\ 0 & -1 & 0 & 0 \\ -1 & 0 & 0 & 0 \end{pmatrix} |\Psi_\eta\rangle$$

$$\gamma^1 \partial_1 |\Psi_1\rangle = i\partial_1 a_{41}$$
$$i\gamma^1 \partial_1 |\Psi_2\rangle = i\partial_1 a_{32}$$
$$i\gamma^1 \partial_1 |\Psi_3\rangle = -i\partial_1 a_{23}$$
$$i\gamma^1 \partial_1 |\Psi_4\rangle = -i\partial_1 a_{14}$$

$$i\gamma^2\partial_2\left|\Psi_\eta\right\rangle = i\partial_2\begin{pmatrix} 0 & 0 & 0 & -i \\ 0 & 0 & i & 0 \\ 0 & i & 0 & 0 \\ -i & 0 & 0 & 0 \end{pmatrix}\left|\Psi_\eta\right\rangle$$

$$i\gamma^2\partial_2\left|\Psi_1\right\rangle = i\left(-i\partial_2 a_{41}\right)$$

$$i\gamma^2\partial_2\left|\Psi_2\right\rangle = i\partial_2 a_{32}$$

$$i\gamma^2\partial_2\left|\Psi_3\right\rangle = i\partial_2 a_{23}$$

$$i\gamma^2\partial_2\left|\Psi_4\right\rangle = i\left(-i\partial_2 a_{14}\right)$$

$$i\gamma^3\partial_3\left|\Psi_\eta\right\rangle = i\partial_3\begin{pmatrix} 0 & 0 & 1 & 0 \\ 0 & 0 & 0 & -1 \\ -1 & 0 & 0 & 0 \\ 0 & 1 & 0 & 0 \end{pmatrix}\left|\Psi_\eta\right\rangle$$

$$i\gamma^3\partial_3\left|\Psi_1\right\rangle = i\partial_3 a_{31}$$

$$i\gamma^3\partial_3\left|\Psi_2\right\rangle = -i\partial_3 a_{42}$$

$$i\gamma^3\partial_3\left|\Psi_3\right\rangle = -i\partial_3 a_{13}$$

$$i\gamma^3\partial_3\left|\Psi_4\right\rangle = i\partial_3 a_{24}$$

For $\left|\Psi_1\right\rangle$, the expression $i\gamma^\nu\partial_\nu\left|\Psi_\eta\right\rangle$ can be written as:

$$i\left(\gamma^0\partial_0 + \gamma^1\partial_1 + \gamma^2\partial_2 + \gamma^3\partial_3\right)\left|\Psi_1\right\rangle = i\begin{pmatrix}\partial_0 & 0 & \partial_3 & \left(\partial_1 - i\partial_2\right)\end{pmatrix}\begin{pmatrix} a_{11} \\ a_{21} \\ a_{31} \\ a_{41} \end{pmatrix}$$

The equation $i\gamma^\nu\partial_\nu\left|\Psi_\eta\right\rangle = m\left|\Psi_\eta\right\rangle$ $\eta = 1,2,3,4$ and $\nu = 0,1,2,3$ is expressed by the following four equations by letting: $\left|\Psi_\eta\right\rangle = \begin{pmatrix} a_{1\eta} \\ a_{2\eta} \\ a_{3\eta} \\ a_{4\eta} \end{pmatrix}$

$$i\begin{pmatrix}\partial_0 & 0 & \partial_3 & (\partial_1 - i\partial_2)\end{pmatrix}\begin{pmatrix}a_{11}\\a_{21}\\a_{31}\\a_{41}\end{pmatrix} = ma_{11}$$

$$i\begin{pmatrix}0 & \partial_0 & (\partial_1 + i\partial_2) & -\partial_3\end{pmatrix}\begin{pmatrix}a_{12}\\a_{22}\\a_{32}\\a_{42}\end{pmatrix} = ma_{22}$$

$$i\begin{pmatrix}-\partial_3 & -\partial_1 + i\partial_2 & -\partial_0 & 0\end{pmatrix}\begin{pmatrix}a_{13}\\a_{23}\\a_{33}\\a_{43}\end{pmatrix} = ma_{33}$$

$$i\begin{pmatrix}-\partial_1 - i\partial_2 & \partial_3 & 0 & -\partial_0\end{pmatrix}\begin{pmatrix}a_{14}\\a_{24}\\a_{34}\\a_{44}\end{pmatrix} = ma_{44}$$

This system of equations can also be seen as a search for matrix diagonalization, where the eigenvalues are equal to m.

2.2.2.6. *Answer to question 6*

Four eigenvectors $|\Psi^{\mu}\rangle$, $\mu = 1, 2, 3, 4$ are solutions of this matrix system. The solutions sought are plane waves, such as kets expressed as: $|\Psi^{\mu}\rangle = u_{\mu}(E, \vec{p})e^{i(\vec{p}.\vec{r}-Et)}$, where $u(E, p)$ is a four component spinor.

For a particle at rest: $\vec{p} = 0$, the spinor is determined by the equation:

$$i\left(\gamma^0\partial_0 + \gamma^1\partial_1 + \gamma^2\partial_2 + \gamma^3\partial_3\right)|\Psi^{\mu}\rangle = i\gamma^0\partial_0|\Psi^{\mu}\rangle = m|\Psi^{\mu}\rangle$$

Since $\partial_0|\Psi^\mu\rangle = -iE|\Psi^\mu\rangle$ and $|\Psi^\mu\rangle = u(E,0)e^{-iEt}$, u is the solution of the equation $E\gamma^0 u = mu$.

As u is a four-component vector in spinor space: $u = \begin{pmatrix} a_1 \\ a_2 \\ a_3 \\ a_4 \end{pmatrix}$, the equation $E\gamma^0 u = mu$ becomes:

$$E\begin{pmatrix} 1 & 0 & 0 & 0 \\ 0 & 1 & 0 & 0 \\ 0 & 0 & -1 & 0 \\ 0 & 0 & 0 & -1 \end{pmatrix}\begin{pmatrix} a_1 \\ a_2 \\ a_3 \\ a_4 \end{pmatrix} = m\begin{pmatrix} a_1 \\ a_2 \\ a_3 \\ a_4 \end{pmatrix}$$

The four solutions are written as:

$$u_1(E,0) = \begin{pmatrix} 1 \\ 0 \\ 0 \\ 0 \end{pmatrix} \rightarrow E = m \qquad u_2(E,0) = \begin{pmatrix} 0 \\ 1 \\ 0 \\ 0 \end{pmatrix} \Rightarrow E = m$$

$$u_3(E,0) = \begin{pmatrix} 0 \\ 0 \\ 1 \\ 0 \end{pmatrix} \rightarrow E = -m \qquad u_4(E,0) = \begin{pmatrix} 0 \\ 0 \\ 0 \\ 1 \end{pmatrix} \rightarrow E = -m$$

If the wave functions of a particle at rest are plane waves, they are described by the equation $|\Psi^k\rangle = u_k(p)e^{-i(Et-\vec{p}\underline{r})}$.

As $p = 0$, the wave functions are written as: $|\Psi^k\rangle = u_k(p)e^{-i(Et)}$

$$|\Psi^1\rangle = u_1 e^{-imt} = \begin{pmatrix} 1 \\ 0 \\ 0 \\ 0 \end{pmatrix} e^{-imt} \quad |\Psi^2\rangle = u_2 e^{-imt} = \begin{pmatrix} 0 \\ 1 \\ 0 \\ 0 \end{pmatrix} e^{-imt}$$

$$|\Psi^3\rangle = u_3 e^{+imt} = \begin{pmatrix} 0 \\ 0 \\ 1 \\ 0 \end{pmatrix} e^{+imt} \quad |\Psi^4\rangle = u_4 e^{+imt} = \begin{pmatrix} 0 \\ 0 \\ 0 \\ 1 \end{pmatrix} e^{+imt}$$

The wave functions $|\Psi^1\rangle$ and $|\Psi^2\rangle$ are described by a vector with four components of the spinor space and correspond to two states of positive energy $E = m$.

It can be verified that the eigenvector $u_1 = \begin{pmatrix} 1 \\ 0 \\ 0 \\ 0 \end{pmatrix}$ is also an eigenvector of the spin matrix $S_z = \begin{pmatrix} \sigma_z & 0 \\ 0 & \sigma_z \end{pmatrix} = \begin{pmatrix} 1 & 0 & 0 & 0 \\ 0 & -1 & 0 & 0 \\ 0 & 0 & 1 & 0 \\ 0 & 0 & 0 & -1 \end{pmatrix}$.

Since $S_z u_1 = S_z \begin{pmatrix} 1 \\ 0 \\ 0 \\ 0 \end{pmatrix} = \frac{1}{2} \begin{pmatrix} 1 & 0 & 0 & 0 \\ 0 & -1 & 0 & 0 \\ 0 & 0 & 1 & 0 \\ 0 & 0 & 0 & -1 \end{pmatrix} \begin{pmatrix} 1 \\ 0 \\ 0 \\ 0 \end{pmatrix} = \frac{1}{2} \begin{pmatrix} 1 \\ 0 \\ 0 \\ 0 \end{pmatrix} = \frac{1}{2} u_1$

where u_1 corresponds to the spin $\frac{1}{2}$.

$$\text{As } S_z u_2 = S_z \begin{pmatrix} 0 \\ 1 \\ 0 \\ 0 \end{pmatrix} = \frac{1}{2} \begin{pmatrix} 1 & 0 & 0 & 0 \\ 0 & -1 & 0 & 0 \\ 0 & 0 & 1 & 0 \\ 0 & 0 & 0 & -1 \end{pmatrix} \begin{pmatrix} 0 \\ 1 \\ 0 \\ 0 \end{pmatrix} = -\frac{1}{2} \begin{pmatrix} 0 \\ 1 \\ 0 \\ 0 \end{pmatrix} = -\frac{1}{2} u_2$$

where u_2 corresponds to the spin $-\frac{1}{2}$.

The wave functions $|\Psi^3\rangle$ and $|\Psi^4\rangle$ correspond to two states of negative energy $E = -m$.

$$\text{As } S_z u_3 = S_z \begin{pmatrix} 0 \\ 0 \\ 1 \\ 0 \end{pmatrix} = \frac{1}{2} \begin{pmatrix} 1 & 0 & 0 & 0 \\ 0 & -1 & 0 & 0 \\ 0 & 0 & 1 & 0 \\ 0 & 0 & 0 & -1 \end{pmatrix} \begin{pmatrix} 0 \\ 0 \\ 1 \\ 0 \end{pmatrix} = \frac{1}{2} \begin{pmatrix} 0 \\ 0 \\ 1 \\ 0 \end{pmatrix} = \frac{1}{2} u_3$$

where u_3 and $|\Psi^3\rangle$ correspond to the spin $\frac{1}{2}$.

$$\text{As } S_z u_4 = S_z \begin{pmatrix} 0 \\ 1 \\ 0 \\ 0 \end{pmatrix} = \frac{1}{2} \begin{pmatrix} 1 & 0 & 0 & 0 \\ 0 & -1 & 0 & 0 \\ 0 & 0 & 1 & 0 \\ 0 & 0 & 0 & -1 \end{pmatrix} \begin{pmatrix} 0 \\ 0 \\ 0 \\ 1 \end{pmatrix} = -\frac{1}{2} \begin{pmatrix} 0 \\ 0 \\ 0 \\ 1 \end{pmatrix} = -\frac{1}{2} u_4$$

where u_4 and $|\Psi^4\rangle$ therefore correspond to the spin $-\frac{1}{2}$.

2.2.2.7. *Answer to question 7*

If $|\Psi\rangle$ is the solution of the Dirac equation of an isolated moving particle, then $H^2|\Psi\rangle = 0$ and:

$$i\hbar \frac{\partial}{\partial t} |\Psi\rangle = H|\Psi\rangle = (\alpha.p + \beta m)|\Psi\rangle$$

By letting $\alpha_1 = \rho_1\sigma_1, \alpha_2 = \rho_1\sigma_2, \alpha_3 = \rho_1\sigma_3, \beta = \rho_3$ H can be written as:

$$H = p_0 + \rho_1\left(\sigma_1 p_1 + \sigma_2 p_2 + \sigma_3 p_3\right) + \rho_3 m = p_0 + \rho_1\left(\vec{\sigma}.\vec{p}\right) + \rho_3 m$$

The expression $\frac{\vec{\sigma}.\vec{p}}{2p}$ represents the spin. If the considered particle is an electron, then $\frac{\vec{\sigma}.\vec{p}}{2p} = \frac{1}{2}\sigma_z$ and the spin has two possible values: 1/2 and -1/2.

The component $\frac{\vec{\sigma}.\vec{p}}{2p}$ of the spin following the direction of the momentum switches with H. The other components of the spin do not switch with H.

As the eigen solutions are common to H and $\frac{\vec{\sigma}.\vec{p}}{2p}$, we obtain for the case of an isolated moving electron four pairs of eigenvalues:

$$\left(E = \sqrt{p^2 + m^2}; \frac{1}{2}\right), \left(E = \sqrt{p^2 + m^2}; -\frac{1}{2}\right),$$

$$\left(E = -\sqrt{p^2 + m^2}; \frac{1}{2}\right), \left(E = -\sqrt{p^2 + m^2}; -\frac{1}{2}\right)$$

To find the solutions of the Dirac equation of an isolated moving particle corresponding to plane waves, 4D wave vectors of the space of the spinors are considered of the form $|\Psi^\mu\rangle = u_\mu(E, \vec{p})e^{i(\vec{p}.\vec{r} - Et)}$.

The Dirac equation of an isolated moving particle is:

$$i\left(\gamma^0\partial_0 + \gamma^1\partial_1 + \gamma^2\partial_2 + \gamma^3\partial_3\right)|\Psi^\mu\rangle = m|\Psi^\mu\rangle$$

As $\partial_0|\Psi^\mu\rangle = -iE|\Psi^\mu\rangle$ and $\partial_k|\Psi^\mu\rangle = ip_k|\Psi^\mu\rangle$ $k = 1,2,2$

Functions $u_\mu(E,\vec{p})$ are the solution of the equation:

$$\left(i\left(\gamma^0\partial_0+\gamma^1\partial_1+\gamma^2\partial_2+\gamma^3\partial_3\right)-m\right)u_\mu(E,\vec{p})=$$
$$\left(E\gamma^0-p_1\gamma^1-p_2\gamma^2-p_2\gamma^3-m\right)u_\mu(E,\vec{p})=0$$

As the matrices γ^k can be written as:

$$\gamma^0=\begin{pmatrix}I & 0\\ 0 & -I\end{pmatrix},\ \gamma^1=\begin{pmatrix}0 & \sigma_1\\ -\sigma_1 & 0\end{pmatrix},\ \gamma^2=\begin{pmatrix}0 & \sigma_2\\ -\sigma_2 & 0\end{pmatrix},\ \gamma^3=\begin{pmatrix}0 & \sigma_3\\ -\sigma_3 & 0\end{pmatrix}$$

The equation $\left(E\gamma^0-p_1\gamma^1-p_2\gamma^2-p_2\gamma^3-m\right)u_\mu(E,\vec{p})=0$ is given by:

$$\left(E\begin{pmatrix}I & 0\\ 0 & -I\end{pmatrix}-p_1\begin{pmatrix}0 & \sigma_1\\ -\sigma_1 & 0\end{pmatrix}-p_2\begin{pmatrix}0 & \sigma_2\\ -\sigma_2 & 0\end{pmatrix}-p_3\begin{pmatrix}0 & \sigma_3\\ -\sigma_3 & 0\end{pmatrix}-m\begin{pmatrix}I & 0\\ 0 & I\end{pmatrix}\right)$$
$$u_\mu(E,\vec{p})=0$$

$$\left(E\begin{pmatrix}I & 0\\ 0 & -I\end{pmatrix}-\begin{pmatrix}0 & \vec{\sigma}\\ -\vec{\sigma} & 0\end{pmatrix}\vec{p}-m\begin{pmatrix}I & 0\\ 0 & I\end{pmatrix}\right)u_\mu(E,\vec{p})=0$$

$$\begin{pmatrix}(E-m)I & -\vec{\sigma}.\vec{p}\\ \vec{\sigma}.\vec{p} & -(E+m)I\end{pmatrix}u_\mu(E,\vec{p})=0$$

$\begin{pmatrix}(E-m)I & -\vec{\sigma}.\vec{p}\\ \vec{\sigma}.\vec{p} & -(E+m)I\end{pmatrix}$ is a 4 x 4 matrix, which is written by four 2 x 2 matrices, $u_\mu(E,\vec{p})$ is a vector of spinor space that has four components that can be written as $u_\mu(E,\vec{p})=\begin{pmatrix}u_A\\ u_B\end{pmatrix}$, and u_A and u_B each represent two components.

The matrix equation becomes:

$$\begin{pmatrix} (E-m)I & -\vec{\sigma}.\vec{p} \\ \vec{\sigma}.\vec{p} & -(E+m)I \end{pmatrix}\begin{pmatrix} u_A \\ u_B \end{pmatrix} = 0$$

thus:

$$\begin{aligned} &(E-m)u_A - (\vec{\sigma}.\vec{p})u_B = 0 \\ &-(E+m)u_B + (\vec{\sigma}.\vec{p})u_A = 0 \end{aligned}$$

$$(\vec{\sigma}.\vec{p}) = \sigma_1 p_1 + \sigma_2 p_2 + \sigma_3 p_3 = \begin{pmatrix} 0 & 1 \\ 1 & 0 \end{pmatrix} p_1 + \begin{pmatrix} 0 & -i \\ i & 0 \end{pmatrix} p_2 + \begin{pmatrix} 1 & 0 \\ 0 & -1 \end{pmatrix} p_3$$

$$(\vec{\sigma}.\vec{p}) = \begin{pmatrix} p_3 & p_1 - ip_2 \\ p_1 + ip_2 & -p_3 \end{pmatrix}$$

Consequently:

$$u_B = \frac{(\vec{\sigma}.\vec{p})}{(E+m)} u_A = \frac{1}{(E+m)} \begin{pmatrix} p_3 & p_1 - ip_2 \\ p_1 + ip_2 & -p_3 \end{pmatrix} u_A$$

If by arbitrary choice $u_A = \begin{pmatrix} 1 \\ 0 \end{pmatrix}$ and $u_A = \begin{pmatrix} 0 \\ 1 \end{pmatrix}$ are set, then the following solutions $u_1(E,\vec{p})$ and $u_2(E,\vec{p})$ are obtained:

$$u_1(E,\vec{p}) = \begin{pmatrix} u_A \\ u_B \end{pmatrix} = \begin{pmatrix} 1 \\ 0 \\ \dfrac{p_3}{E+m} \\ \dfrac{p_1 + ip_2}{E+m} \end{pmatrix} \text{ and } u_2(E,\vec{p}) = \begin{pmatrix} u_A \\ u_B \end{pmatrix} = \begin{pmatrix} 0 \\ 1 \\ \dfrac{p_1 - ip_2}{E+m} \\ \dfrac{-p_3}{E+m} \end{pmatrix}$$

If by arbitrary choice $u_B = \begin{pmatrix} 1 \\ 0 \end{pmatrix}$ and $u_B = \begin{pmatrix} 0 \\ 1 \end{pmatrix}$ are set, the following solutions $u_3(E,\vec{p})$ and $u_4(E,\vec{p})$ are obtained:

$$u_3(E,\vec{p}) = \begin{pmatrix} u_A \\ u_B \end{pmatrix} = \begin{pmatrix} \dfrac{p_3}{E-m} \\ \dfrac{p_1 + ip_2}{E-m} \\ 1 \\ 0 \end{pmatrix} \text{ and } u_4(E,\vec{p}) = \begin{pmatrix} u_A \\ u_B \end{pmatrix} = \begin{pmatrix} \dfrac{p_1 - ip_2}{E-m} \\ \dfrac{-p_3}{E-m} \\ 0 \\ 1 \end{pmatrix}$$

The Dirac equation therefore has:

– two positive energy solutions

$$\left|\Psi^1\right\rangle = u_1\left(E,p\right)e^{-i(Et-\vec{p}.\vec{r})} = N_a \begin{pmatrix} 1 \\ 0 \\ \dfrac{p^3}{E+m} \\ \dfrac{p^1 + ip^2}{E+m} \end{pmatrix} e^{-i(Et-\vec{p}.\vec{r})} \; u_1(E,\vec{p}) \text{ t}$$

$$\left|\Psi^2\right\rangle = u_2\left(E,p\right)e^{-i(Et-\vec{p}.\vec{r})} = N_a \begin{pmatrix} 0 \\ 1 \\ \dfrac{p^1 - ip^2}{E+m} \\ \dfrac{-p^3}{E+m} \end{pmatrix} e^{-i(Et-\vec{p}.\vec{r})}$$

where N_a is a normalization coefficient equal to $\sqrt{\dfrac{2E}{E+m}}$;

– two negative energy solutions: $u_3(E,\vec{p})$ and $u_4(E,\vec{p})$

$$\left|\Psi^3\right\rangle = N_b \begin{pmatrix} \dfrac{p^3}{E-m} \\ \dfrac{p^1+ip^2}{E-m} \\ 1 \\ 0 \end{pmatrix} e^{-i(Et-\vec{p}.\vec{r})}$$

$$\left|\Psi^4\right\rangle = N_b \begin{pmatrix} \dfrac{p^1-ip^2}{E-m} \\ \dfrac{-p^3}{E-m} \\ 1 \\ 0 \end{pmatrix} e^{-i(Et-\vec{p}.\vec{r})}$$

where N_b is a normalization coefficient that is set equal to $\sqrt{\dfrac{2E}{E-m}}$.

2.2.2.8. *Answer to question 8*

The ket $\left|\Psi^1\right\rangle$ is the solution of the Dirac equation if the operator A is the same as the operator B. Operators A and B are defined by:

$$A\left|\Psi^1\right\rangle = \sqrt{\frac{2E}{E+m}}\left(i\left(\partial_0 a_{11} + \partial_3 a_{31} + \left(\partial_1 - i\partial_2\right)a_{41}\right)\right)$$

And $B\left|\Psi^1\right\rangle = \sqrt{\dfrac{2E}{E+m}}\left(ma_{11}\right)$

As $a_{11}=1$,

$$B = \sqrt{\frac{2E}{E+m}}(m)$$

If $\widehat{p}$ is the momentum operator and E is the energy operator, then

$$E^2 = \widehat{p}^2 + m^2$$

$$\partial_0 \to \frac{E}{i}$$

$$\frac{\partial}{\partial x} \to i\widehat{p}$$

$$A = i\sqrt{\frac{2E}{E+m}}\left(\partial_0 + \partial_3 \frac{p^3}{E+m} + (\partial_1 - i\partial_2)\left(\frac{p^1 + ip^2}{E+m}\right)\right)$$

$$A = i\sqrt{\frac{2E}{E+m}}\left(\partial_0 + \frac{1}{E+m}\left(\partial_3 p^3 + \partial_1 p^1 + \partial_2 p^2\right)\right)$$

$$A = i\sqrt{\frac{2E}{E+m}}\left(\frac{E}{i} + \frac{i\widehat{p}^2}{E+m}\right) = \sqrt{\frac{2E}{E+m}}\left(E - \frac{\widehat{p}^2}{E+m}\right)$$

$$A = \sqrt{\frac{2E}{E+m}}\left(\frac{E^2 + mE - \widehat{p}^2}{E+m}\right) = \sqrt{\frac{2E}{E+m}}\left(\frac{E^2 + mE - \left(E^2 - m^2\right)}{E+m}\right)$$

$$A = \sqrt{\frac{2E}{E+m}}\left(\frac{m\left(E+m\right)}{E+m}\right) = \sqrt{\frac{2E}{E+m}}(m)$$

The operators A and B are identical and $\left|\Psi^1\right\rangle$ is the solution of the Dirac equation.

To verify that ket $\left|\Psi^2\right\rangle$ is a solution of the system of equations, the operators C and D are considered. These operators are defined by:

$$C\left|\Psi^2\right\rangle = \sqrt{\frac{2E}{E+m}} i \begin{pmatrix} 0 & \partial_0 & (\partial_1 + i\partial_2) & -\partial_3 \end{pmatrix} \begin{pmatrix} a_{12} \\ a_{22} \\ a_{32} \\ a_{42} \end{pmatrix}$$

$$C\left|\Psi^2\right\rangle = \sqrt{\frac{2E}{E+m}} \left(i\left(\partial_0 a_{22} + (\partial_1 + i\partial_2) a_{32} - \partial_3 a_{42} \right) \right)$$

And $D\left|\Psi^2\right\rangle = \sqrt{\frac{2E}{E+m}} (m a_{22})$

As $a_{22} = 1$

$$D = \sqrt{\frac{2E}{E+m}} (m)$$

$$C = \sqrt{\frac{2E}{E+m}} \left(i\left(\frac{E}{i} + \frac{(\partial_1 + i\partial_2)(p_1 - ip_2)}{E+m} + \frac{\partial_3 p_3}{E+m} \right) \right)$$

$$C = \sqrt{\frac{2E}{E+m}} \left(i\left(\frac{E}{i} + \frac{\partial_1 p_1 + \partial_2 p_2 + \partial_3 p_3}{E+m} \right) \right)$$

$$C = \sqrt{\frac{2E}{E+m}} \left(i\left(\frac{E}{i} + \frac{\partial_1 p_1 + \partial_2 p_2 + \partial_3 p_3}{E+m} \right) \right) =$$

$$C = \sqrt{\frac{2E}{E+m}} \left(i\left(\frac{E}{i} + \frac{ip^2}{E+m} \right) \right)$$

thus: $C = \sqrt{\frac{2E}{E+m}} (m) = D$

The operators C and D are identical.

To verify that ket $|\Psi^3\rangle$ is a solution of the system of equations, the operators K and L are considered. These operators are defined by:

$$K|\Psi^3\rangle = \sqrt{\frac{2E}{E-m}} i\begin{pmatrix} -\partial_3 & -\partial_1 + i\partial_2 & -\partial_0 & 0 \end{pmatrix}\begin{pmatrix} a_{13} \\ a_{23} \\ a_{33} \\ a_{43} \end{pmatrix}$$

$$K = \sqrt{\frac{2E}{E-m}}\left(i\left(\partial_0 a_{33} - (\partial_1 - i\partial_2)a_{23} - \partial_3 a_{13}\right)\right)$$

and $L|\Psi^3\rangle = \sqrt{\frac{2E}{E-m}}(ma_{33}) = \sqrt{\frac{2E}{E-m}}m$

$$K = \sqrt{\frac{2E}{E-m}}\left(i\left(-\partial_0 - (\partial_1 - i\partial_2)\frac{(p_1 + ip_2)}{E-m} - \partial_3 \frac{p_3}{E-m}\right)\right)$$

$$K = \sqrt{\frac{2E}{E-m}}\left(i\left(-\frac{E}{i} - (\partial_1 p_1 + \partial_2 p_2 + \partial_3 p_3)\right)\right)$$

$$K = \sqrt{\frac{2E}{E-m}}\left(i\left(-\frac{E}{i} - \frac{ip^2}{E-m}\right)\right)$$

$$K == \sqrt{\frac{2E}{E-m}}\left(\frac{-E(E-m) + p^2}{E-m}\right) = \sqrt{\frac{2E}{E-m}}\left(\frac{-E^2 + (Em) + E^2 - m}{E-m}\right)$$

So $K == \sqrt{\frac{2E}{E-m}}\left(\frac{-E(E-m) + p^2}{E-m}\right) = \sqrt{\frac{2E}{E-m}}m = L$

The operations K and L are identical.

To verify that ket $\left|\Psi^4\right\rangle$ is a solution of the system of equations, the operators M and N are considered. These operators are defined by

$$M\left|\Psi^4\right\rangle = \sqrt{\frac{2E}{E-m}}\,i\left(-\partial_1 - i\partial_2 \quad \partial_3 \quad 0 \quad -\partial_0\right)\begin{pmatrix} a_{14} \\ a_{24} \\ a_{34} \\ a_{44} \end{pmatrix}$$

$$M\left|\Psi^3\right\rangle = \sqrt{\frac{2E}{E-m}}\,i\left(-\partial_0 + \left(-\partial_1 - i\partial_2\right)\frac{\left(p_1 - ip_2\right)}{E-m} - \partial_3\frac{p_3}{E-m}\right)$$

$$M = \sqrt{\frac{2E}{E-m}}\,i\left(-\frac{E}{i} - \frac{\partial_1 p_1 + \partial_2 p_2 + \partial_3 p_3}{E-m}\right)$$

$$M = \sqrt{\frac{2E}{E-m}}\left(i\left(-\frac{E}{i} - \frac{ip^2}{E-m}\right)\right)$$

$$N\left|\Psi^4\right\rangle = \sqrt{\frac{2E}{E-m}}\left(ma_{44}\right) \Rightarrow N = \sqrt{\frac{2E}{E-m}}\left(m\right)$$

so $M = N = \sqrt{\frac{2E}{E-m}}\left(m\right)$

The operators M and N are identical.

2.2.2.9. *Answer to question 9*

According to the Stückelberg–Feynman interpretation, a negative energy state corresponds to an antiparticle state by reversing E and P:

$$v_1(E,\vec{p}) = u_4(-E,-\vec{p})$$

$$v_2(E,\vec{p}) = u_3(-E,-\vec{p})$$

Since the proper time of the antiparticles passes in the reverse direction of time, the antiparticles have positive energy.

The expressions of the kets solutions of the Dirac equation of an isolated moving particle are:

$$\left|\Psi^1\right\rangle = u_1(E,p)e^{-i(Et-\vec{p}.\vec{r})} = \sqrt{\frac{2E}{E+m}}\begin{pmatrix} 1 \\ 0 \\ \dfrac{p^3}{E+m} \\ \dfrac{p^1+ip^2}{E+m} \end{pmatrix} e^{-i(Et-\vec{p}.\vec{r})}$$

The ket $\left|\Psi^1\right\rangle$ is the product of a 4D spinor in spinor space and a plane wave defined by $e^{-i(Et-\vec{p}.\vec{r})}$ in physical space. In the case of the electron, the wave function $\left|\Psi^1\right\rangle$ corresponds to a positive energy $\left(E=\sqrt{p^2+m^2}\right)$ and a spin $\frac{1}{2}$:

$$\left|\Psi^2\right\rangle = u_2(E,p)e^{-i(Et-\vec{p}.\vec{r})} = \sqrt{\frac{2E}{E+m}}\begin{pmatrix} 0 \\ 1 \\ \dfrac{p^1-ip^2}{E+m} \\ \dfrac{-p^3}{E+m} \end{pmatrix} e^{-i(Et-\vec{p}.\vec{r})}$$

The ket $\left|\Psi^2\right\rangle$ is the product of a 4D spinor in spinor space and a plane wave defined by e^{-ipr} in physical space. In the case of the electron, the wave function corresponds $\left|\Psi^2\right\rangle$ to a particle of positive energy $E=\sqrt{p^2+m^2}$ and spin $-\frac{1}{2}$. The ket:

$$\left|\Psi^3\right\rangle = v_2(E,p)e^{-i(Et-\vec{p}.\vec{r})} = u_3(-E,-p)e^{-i(Et-\vec{p}.\vec{r})}$$

$$\left|\Psi^3\right\rangle = v_2(E,p) = \sqrt{\frac{-2E}{-E-m}} \begin{pmatrix} \frac{p^3}{E+m} \\ \frac{p^1+ip^2}{E+m} \\ 1 \\ 0 \end{pmatrix} e^{-i(Et-\vec{p}.\vec{r})}$$

$$\left|\Psi^3\right\rangle = \sqrt{\frac{2E}{E+m}} \begin{pmatrix} \frac{p^3}{E+m} \\ \frac{p^1+ip^2}{E+m} \\ 1 \\ 0 \end{pmatrix} e^{-i(Et-\vec{p}.\vec{r})}$$

is the product of a 4D spinor in spinor space and a plane wave defined by $e^{-i(Et-\vec{p}.\vec{r})}$ in physical space. In the case of the electron, the wave function $\left|\Psi^3\right\rangle$ corresponds to an antiparticle of energy $\left(E=\sqrt{p^2+m^2}\right)$ and spin $\frac{1}{2}$:

The ket

$$\left|\Psi^4\right\rangle = v_1(-E,-p)e^{-i(Et-\vec{p}.\vec{r})} = u_4(-E,-p)e^{-i(Et-\vec{p}.\vec{r})} = \sqrt{\frac{2E}{E+m}} \begin{pmatrix} \frac{p^1-ip^2}{E-m} \\ \frac{-p^3}{E-m} \\ 0 \\ 1 \end{pmatrix} e^{-i(Et-\vec{p}.\vec{r})}$$

is the product of a 4D spinor in spinor space and a plane wave defined by e^{ipr} in physical space. In the case of the electron, the wave function $\left|\Psi^4\right\rangle$ corresponds to an antiparticle of energy $\left(E=\sqrt{p^2+m^2}\right)$ and spin $-\frac{1}{2}$.

The most general form of the solution to Dirac's equation is a sum weighted by complex coefficients of all eigenstates.

2.2.2.10. *Answer to question 10*

The dot product of the spinor with itself is written as:

$$u_1^\dagger u_1 = \frac{E+m}{2E}\begin{pmatrix} 1 & 0 & \dfrac{p_3}{E+m} & \dfrac{p_1 - ip_2}{E+m} \end{pmatrix}\begin{pmatrix} 1 \\ 0 \\ \dfrac{p_3}{E+m} \\ \dfrac{p_1 + ip_2}{E+m} \end{pmatrix}$$

$$u_1^\dagger u_1 = \frac{E+m}{2E}\left(1 + \frac{p^2}{(E+m)^2}\right) = \frac{E+m}{2E}\left(\frac{E^2 + m^2 + 2Em + p^2}{(E+m)^2}\right)$$

$$u_1^\dagger u_1 = \frac{E+m}{2E}\left(\frac{2E(E+m)}{(E+m)^2}\right) = \frac{E+m}{2E}\left(\frac{2E}{E+m}\right) = 1$$

The dot product of the spinor u_2 with itself is given by:

$$u_2^\dagger u_2 = \frac{E+m}{2E}\begin{pmatrix} 0 & 1 & \dfrac{p_1 + ip_2}{E+m} & \dfrac{-p_3}{E+m} \end{pmatrix}\begin{pmatrix} 0 \\ 1 \\ \dfrac{p_1 - ip_2}{E+m} \\ \dfrac{-p_3}{E+m} \end{pmatrix}$$

$$u_2^\dagger u_2 = \frac{E+m}{2E}\left(1 + \frac{p^2}{(E+m)^2}\right) = 1$$

The dot product of the spinor v_2 with itself is written as:

$$v_2^\dagger v_2 = \frac{E+m}{2E}\begin{pmatrix} \frac{p_3}{E+m} & \frac{p_1 - ip_2}{E+m} & 1 & 0 \end{pmatrix}\begin{pmatrix} \frac{p_3}{E+m} \\ \frac{p_1 + ip_2}{E+m} 0 \\ 1 \\ 0 \end{pmatrix}$$

$$v_2^\dagger v_2 = \frac{E+m}{2E}\left(1 + \frac{p^2}{(E+m)^2}\right) = 1$$

The dot product of the spinor v_1 with itself is written as:

$$v_1^\dagger v_1 = \frac{E+m}{2E}\begin{pmatrix} \frac{p_1 + ip_2}{E+m} & \frac{-p_3}{E+m} & 0 & 1 \end{pmatrix}\begin{pmatrix} \frac{p_1 - ip_2}{E+m} \\ \frac{-p_3}{E+m} \\ 0 \\ 1 \end{pmatrix}$$

$$v_1^\dagger v_1 = \frac{E+m}{2E}\left(1 + \frac{p^2}{(E+m)^2}\right) = 1$$

2.2.2.11. *Answer to question 11*

To verify that the spinors u_1 and u_2 are orthogonal, the dot product of the spinor u_1 with u_2 is calculated as:

$$u_1^\dagger u_2 = \frac{E+m}{2E}\begin{pmatrix} 1 & 0 & \frac{p_3}{E+m} & \frac{p_1 - ip_2}{E+m} \end{pmatrix}\begin{pmatrix} 1 \\ 0 \\ \frac{p_1 - ip_2}{E+m} \\ \frac{-p_3}{E+m} \end{pmatrix} = \frac{E+m}{2E}(0) = 0$$

The dot product of the spinor u_2 with u_1 is written as:

$$u_2^\dagger u_1 = \frac{E+m}{2E}\begin{pmatrix} 0 & 1 & \frac{p_1+ip_2}{E+m} & \frac{-p_3}{E+m} \end{pmatrix}\begin{pmatrix} 1 \\ 0 \\ \frac{p_3}{E+m} \\ \frac{p_1+ip_2}{E+m} \end{pmatrix} = \frac{E+m}{2E}(0) = 0$$

To verify that the spinors v_1 and v_2 are orthogonal, the dot product of the spinor v_1 with v_2 is calculated as:

$$v_1^\dagger v_2 = \frac{E+m}{2E}\begin{pmatrix} \frac{p_1+ip_2}{E+m} & \frac{-p_3}{E+m} & 0 & 1 \end{pmatrix}\begin{pmatrix} \frac{p_3}{E+m} \\ \frac{p_1+ip_2}{E+m} \\ 1 \\ 0 \end{pmatrix} = \frac{E+m}{2E}(0) = 0$$

The dot product of the spinor v_2 with v_1 is written as:

$$v_2^\dagger v_1 = \frac{E+m}{2E}\begin{pmatrix} \frac{p_3}{E+m} & \frac{p_1-ip_2}{E+m} & 1 & 0 \end{pmatrix}\begin{pmatrix} \frac{p_1-ip_2}{E+m} \\ \frac{-p_3}{E+m} \\ 1 \\ 0 \end{pmatrix} = \frac{E+m}{2E}(0) = 0$$

To verify that the spinors $u_1(p)$ and $v_2(-p)$ are orthogonal, the dot product of the spinor $u_1(p)$ with the spinor $v_2(-p)$ is calculated as:

$$u_1^\dagger(p)u_3(-p)=\frac{\sqrt{(E+m)(E+m)}}{2E}\begin{pmatrix}1 & 0 & \frac{p_3}{E+m} & \frac{p_1-ip_2}{E+m}\end{pmatrix}\begin{pmatrix}-\frac{p_3}{E+m}\\ -\frac{p_1+ip_2}{E+m}\\ 1\\ 0\end{pmatrix}$$

$$u_1^\dagger(p)v_2(-p)=\frac{\sqrt{(E+m)(E+m)}}{2E}(0)=0$$

To verify that the spinors $u_1(p)$ and $v_1(-p)$ are orthogonal, the dot product of the spinor $u_1(p)$ with $v_1(-p)$ the spinor is given by:

$$u_1^\dagger(p)v_1(-p)=\frac{\sqrt{(E+m)(E+m)}}{2E}\begin{pmatrix}1 & 0 & \frac{p_3}{E+m} & \frac{p_1-ip_2}{E+m}\end{pmatrix}\begin{pmatrix}\frac{-(p_1-ip_2)}{E+m}\\ \frac{p_3}{E+m}\\ 0\\ 1\end{pmatrix}$$

$$u_1^\dagger(p)v_1(-p)=\frac{\sqrt{(E+m)(E+m)}}{2E}(0)=0$$

2.3. The density matrix for a two-level laser system

The correct and complete description of the principles by which a laser operates is obtained by coupling Maxwell's equations to quantum mechanics theory. It is then necessary to introduce into the mathematical model, the density matrix of the system under study.

The first part of this exercise deals with the definition of the density matrix. In the second part, the equation of motion of the density matrix is established. In the third part, a two-level system interacting with an electromagnetic field is studied.

2.3.1. *Definition of the density matrix*

The Hamiltonian H_0 of an isolated atomic system is written in terms of kinetic energy and potential energy operators as:

$$H_0 = -\frac{\hbar^2}{2m}\Delta + V(r) \qquad [2.1]$$

where $-\frac{\hbar^2}{2m}\Delta$ is the kinetic energy operator of the system and $V(r)$ is the potential energy operator of the system.

1) Give the expression of the eigenvalue stationary Schrödinger equation in the absence of an external field from which the eigen functions $U_n(r)$ and the eigenvalues $E_n(r)$ of this system are determined.

Solution:

$$H_0 U_n(r) = \left\{-\frac{\hbar^2}{2m}\Delta + V\right\} U_n(r) = E_n(r) U_n(r) \qquad [2.2]$$

2) Deduce, as a function of the eigen functions $U_n(r)$ of the Hamiltonian H_0, the expression of the wave function $\Psi\ (r,\ t)$ describing, at time t, a state of the system in the presence of an electromagnetic field.

Solution:

$$\Psi(r,t) = \sum_n c_n(t) U_n(r) \qquad [2.3]$$

A general quantum state is a linear combination of the eigenstates of the Hamiltonian.

3) The mean value of an operator $\hat{A}$ (observable) is expressed as:

$$\langle \Psi(r,t)|\hat{A}|\Psi(r,t)\rangle = \int \Psi^*(r,t)\hat{A}\Psi(r,t)d\tau \qquad [2.4]$$

where $d\tau = d^3x$ is a volume element in 3D space.

Show that:

$$\langle \hat{A}\rangle_{mean} = \sum_m \sum_n c_m^*(t) A_{mn} c_n(t) \qquad [2.5]$$

where $A_{mn} = \langle \mathrm{U}_m|\hat{A}|\mathrm{U}_n\rangle$

Solution:

The expression of the mean value of an operator $\hat{A}$ is given by:

$$\langle \hat{A} \rangle = \int \left\{ \sum_m c_m^*(t) U_m^*(r)\, \hat{A} \left[\sum_n c_n(t) U_n(r) \right] \right\} d\tau$$

$$= \sum_m \sum_n c_m^*(t) \left[\int U_m^*(r) \hat{A} U_n(r) d\tau \right] c_n(t) \qquad [2.6]$$

where $A_{mn} = \int U_m^*(r) \hat{A} U_n(r) d\tau = \langle \mathrm{U}_m | \hat{A} | \mathrm{U}_n \rangle$.

4) What does the bi-linear product $c_m^*(t) c_n(t)$ represent?

The expression of the mean of operator A shows that the probability of finding the matrix element A_{mn} must be equal to the bi-linear product $c_m^*(t) c_n(t)$.

Since each atom of the system can be disturbed in a different way, there are an infinite number of states $\Psi_s(r,t)$ that can be used to describe each subsystem. The latter is composed of an atom in its environment characterized by index s and which labels one of the possible states. Under these conditions, the mean value $\langle \hat{A} \rangle_s$ is one of the possible values of the measurement of A.

5) Give the expression of the mean ensemble $\overline{\langle \hat{A} \rangle_s}$ (the probability of measuring $\langle \hat{A} \rangle_s$, which is denoted by P_s).

Solution:

By definition: the mean value $\langle\langle \hat{A} \rangle_s$ is given by:

$$\overline{\langle \hat{A} \rangle_s} = \sum_s P_s \langle \hat{A} \rangle_s \qquad [2.7]$$

6) Show that:

$$\overline{\langle A \rangle_s} = \sum_{m,n} \left\{ \sum_s P_s \left[C_m^{*(s)}(t) \right] \left[C_n^{(s)}(t) \right] \right\} A_{mn}$$

$$\overline{\langle \mathrm{A} \rangle_s} = \sum_{m,n} \left\langle \mathrm{C}_{\mathrm{m}}^{*(\mathrm{s})}(t)\mathrm{C}_{\mathrm{n}}^{(\mathrm{s})}(\mathrm{t}) \right\rangle A_{mn} = \sum_{m,n} \rho_{nm} A_{mn}$$

where ρ_{nm} is an element of the density matrix.

Solution:

Starting from $\overline{\langle \hat{A} \rangle_s} = \sum_s P_s \langle \hat{A} \rangle_s$ and expanding [2.6], it is shown that:

$$\overline{\langle \hat{A} \rangle_s} = \sum_s P_s \langle \hat{A} \rangle_s = \sum_m \sum_n \left\{ \sum_s P_s \, c_m^{(s)*}(t) c_n^s(t) \right\} A_{mn}$$

since each atom in its environment s contributes to the general state $\Psi^s(r,t) = \sum_n c_n^s(t) U_n(r)$. Atoms have been grouped into representative classes and, in this equation, the notation with the superscript s in $c_n^s(t)$ indicates that the coupling with the environment is different for each class.

Consequently, by averaging over the coefficients $c_n^s(t)$ such that $\sum_s P_s \, c_m^{(s)*}(t) c_n^s(t) = \rho_{nm}$ (inversion of the indices m and n), it can be written that: $\overline{\langle \hat{A} \rangle_s} = \sum_s P_s \langle \hat{A} \rangle_s = \sum_m \sum_n \rho_{nm} A_{mn}$

where ρ_{nm} is an element of the density matrix.

7) Show that $\langle \hat{A} \rangle = Trace\ (\rho A)$, where the overall mean notation is omitted for simplicity.

Solution:

Using matrix multiplication rules whereby an element jk of the product of two matrices A and B is written as $(AB)_{jk} = \sum_m A_{jm} B_{mk}$, the diagonal element $(AB)_{nn}$ is given by: $(AB)_{nn} = \sum_m A_{nm} B_{mn}$.

The trace being expressed as: $TrAB = \sum_n AB_{nn}$

$$\overline{\langle \hat{A} \rangle_s} = \sum_n \sum_m \rho_{nm} A_{mn} = \sum_n (\rho A)_{nn} = Tr(\rho A)$$

8) Replacing A by the identity operator, show that: trace:

$$Tr(\rho) = \sum_n \rho_{nn} = 1$$

Solution:

Using I instead of A,

$$\langle \hat{I} \rangle = Trace\ (\rho I) = Tr(\rho) = \sum_n \rho_{nn}$$

Because: $\langle \mathrm{U}_m | \hat{I} | \mathrm{U}_n \rangle = \delta_{mn}$, then $\rho_{nn} = \langle c_n^*(t) c_n(t) \rangle$ and $\rho_{mn} = 0$ if $m \neq n$ Given that the eigenvectors $\Psi(r,t) = \sum_n c_n(t) U_n(r)$ are orthonormal: $\sum_n \rho_{nn} = \sum_n |\langle c_n(t) \rangle|^2 = 1$.

9) Show that the density matrix is Hermitian.

Solution:

$$\rho_{nm} = \langle c_m^*(t) c_n(t) \rangle \text{ and } \rho_{nm}^* = \langle c_m^*(t) c_n(t) \rangle^* = \rho_{mn}$$

10) What does $\rho_{nn}N$ represent (where N is the number of atoms comprising the system per unit volume)?

Solution:

$\rho_{nn}N$ represents the number of atoms in state n.

11) Let μ be the dipole moment operator associated with the induced dipole moment (the permanent dipole moment is assumed to be zero), which induces a transition between state n = 1 and state n = 2.

Show that: $\langle \mu \rangle = \mu_{21}(\rho_{21} + \rho_{21}^*)$ and that $\langle \mu \rangle$ is real.

Solution:

By definition:

$$\langle \mu \rangle = \sum_n \sum_m \rho_{nm} \mu_{mn} = \rho_{11}\mu_{11} + \rho_{12}\mu_{21} + \rho_{21}\mu_{12} + \rho_{22}\mu_{22}$$

Since the permanent dipole moment is zero, $\mu_{11} = \mu_{22} = 0$, $\mu_{21} = \mu_{12}$ and that $\rho_{12} = \rho_{21}^*$:

$$\langle\mu\rangle = \mu_{21}(\rho_{21} + \rho_{21}^*).$$

Let $\rho_{21} = |\rho_{21}|e^{i\theta}$;

Then, $\rho_{21} + \rho_{21}^* = |\rho_{21}|e^{i\theta} + |\rho_{21}|e^{-i\theta} = 2|\rho_{21}|cos\theta$; hence, the dipole moment $\langle\mu\rangle$ is a real number.

12) Deduce the induced polarization P (N is the number of atoms per unit of volume).

Solution:

$$P = N\langle\mu\rangle = N\mu_{21}(\rho_{21} + \rho_{21}^*)$$

2.3.2. *Density matrix properties*

It is assumed that atoms are prepared so that each wave function has the same phase origin. In this case: $P_s(\theta_s) = \delta(\theta_s - \theta_0)$, where θ_0 is the phase at the origin, such that: $\int_0^{2\pi} P_s(\theta_s)d\theta_s = 1$. The environment acts identically on every atom.

In the case of a single atom, the wave function is given by:

$$|\Psi_m\rangle = |U_m\rangle e^{-i\omega_m t} e^{-i\theta_m} \quad [2.8]$$

where $\omega_m = \dfrac{E_m}{\hbar}$.

1) Check that $|\Psi_m\rangle$ is the solution of the time-dependent Schrödinger equation.

Solution:

The time-dependent Schrödinger equation has the following expression:

$$H_0\psi_m(r,t) = \left\{-\frac{\hbar^2}{2m}\Delta + V\right\}\psi_m(r,t) = -i\hbar\frac{\partial}{\partial t}\psi_m(r,t) \quad [2.9]$$

with:

$$-i\hbar\frac{\partial}{\partial t}\psi_m(r,t) = i\hbar\frac{\partial}{\partial t}\left(U_m(r)e^{-i\omega_m t}e^{-i\theta_m}\right)$$
$$= \hbar\omega_m U_m(r)e^{-i\omega_m t}e^{-i\theta_m}$$

and

$$H_0\psi_m(r,t) = \left\{-\frac{\hbar^2}{2m}\Delta + V\right\}\left(U_m(r)e^{-i\omega_m t}e^{-i\theta_m}\right)$$
$$= E_m U_m(r)e^{-i\omega_m t}e^{-i\theta_m}$$

Since $E_m = \hbar\omega_m$, $\psi_m(r,t) = U_m(r)e^{-i\omega_m t}e^{-i\theta_m}$ is a solution of the time-dependent Schrödinger equation.

2) Give the expression of ρ_{nm}

Solution:

From the equation $\sum_s P_s\, c_m^{(s)*}(t)c_n^s(t) = \rho_{nm}$, which can be transformed by replacing the sum with an integral:

$$\rho_{nm} = \int_{\theta_s=0}^{2\pi} \delta(\theta_s - \theta_0)\, c_m^{(s)*}(t)e^{+i\theta_m^{(s)}} c_n^s(t)e^{-i\theta_n^{(s)}} e^{i\omega_{mn}t} d\theta_s,$$

such that: $\rho_{nm} = c_m^*(t)c_n\,(t)e^{i\omega_{mn}t} \int_{\theta_s=0}^{2\pi} \delta(\theta_s - \theta_0)\, e^{+i\left(\theta_m^{(s)} - \theta_n^{(s)}\right)} d\theta_s,$

which leads to: $\rho_{nm} = c_m^*(t)c_n\,(t)e^{i\omega_{mn}t}e^{+i\theta_0}$. It must be noted that if at time $t = 0$ the phases are the same, they are in any case different at time t. As a result, state n and state m do not *a priori* have the same phases in the integral, which leads to the term $e^{+i\theta_0}$, which can be assumed to be such that $e^{+i\theta_0} = 1$.

3) Show that ρ_{nn} does not depend on θ_0 and comment on the case of ρ_{nn}?

Solution:

The diagonal terms are not affected by the phase distribution since $e^{+i\left(\theta_n^{(s)} - \theta_n^{(s)}\right)} = 1$ and $\rho_{nn} = c_n^*(t)c_n\,(t) = \left|c_n\,(t)\right|^2$ and we have $\rho_{nn} = \frac{N_n}{N}$.

If it is assumed that the phases of the states of the atoms are uniformly distributed over 2π, then: $P_s(\theta_s) = \frac{1}{2\pi} \quad 0 \leq \theta_s < 2\pi$.

4) Show that $\rho_{nm} = 0$ if $n \neq m$.

Solution:

$$\rho_{nm} = \left(\frac{1}{2\pi} \int_{\theta_s=0}^{2\pi} c_m^{(s)*}(t) e^{+i\theta_m^{(s)}} c_n^s(t) e^{-i\theta_n^{(s)}} d\theta_s \right) e^{i\omega_{mn}t}$$

such that:

$$\rho_{nm} = c_m^{(s)*}(t) c_n^s(t) e^{i\omega_{mn}t} \left(\frac{1}{2\pi} \int_{\theta_s=0}^{2\pi} e^{+i\theta_m^{(s)}} e^{-i\theta_n^{(s)}} d\theta_s \right) = 0$$

since: $\frac{1}{2\pi} \int_{\theta_s=0}^{2\pi} e^{+i\left(\theta_m^{(s)} - \theta_n^{(s)}\right)} d\theta_s = 0$.

5) Analyze the fact that $\rho_{nm} = 0$ if n ≠ m.

Solution:

If the phases are distributed over a period of 2π, all the non-diagonal elements of the density matrix are always zero. This means that there is no polarization because $P = N\langle\mu\rangle = N\mu_{21}(\rho_{21} + \rho_{21}^*) = 0$ and that there is no interaction with the electromagnetic field.

5) What is the effect of any disturbance that randomizes the phase of the wave functions on the polarization?

Solution:

Any effect that makes the phase of one state m random with respect to another state n tends to reduce ρ_{nm} and thus the polarization of the medium. This consequence cannot be derived from the balance equations.

6) Give examples of the cases of the phenomena discussed in question 5 above.

Solution:

In interactions between systems, the coupling results either from elastic collisions or from inelastic collisions. Collisions do not necessarily involve contact situations such as with hard spheres, but can be achieved from a distance. The interaction potential is not infinite at the point of contact and zero elsewhere as for hard spheres, but is represented by a curve given, for example, by 12-6 Lennard-Jones or Buckingham-type functions with a repulsive part and an attractive part.

In an elastic collision, between an atom in a state n and another atom of the same type in a state n or another atom of another type in a state n', if the atom does not change state, the phase of a state m necessarily varies and as the collisions are random, this phase varies randomly too.

2.3.3. *Equation of motion of the density matrix*

By definition of the density matrix, the matrix element ρ_{nm} of this operator is the spatial average of the product $c_m(t)^*c_n(t)$ of the components of the wave function describing the state of the system, such as: $\rho_{nm} = \langle c_m(t)^*c_n(t)\rangle$.

Consider a state of the system given by a linear combination of the eigenstates of the Hamiltonian such that: $\langle r|\Psi\rangle(t) = \Psi(r,t) = \sum_k c_k(t)U_k(r)$

1) Show by applying the Schrödinger equation that the density matrix satisfies the Liouville equation, i.e. $\frac{\partial\rho_{nm}}{\partial t} = \frac{1}{i\hbar}[H,\rho]_{nm}$

where: $[H,\rho]_{nm} = \sum_k c_m^*(t)c_k(t)H_{nk} - \sum_k H_{km}c_k^*(t)c_n(t)$

Solution:

From: $\frac{\partial\rho_{nm}}{\partial t} = \frac{\partial c_m^*(t)}{\partial t}c_n(t) + c_m^*(t)\frac{\partial c_n(t)}{\partial t}$

and: $$\frac{\partial\Psi(r)}{\partial t} = \sum_k \frac{\partial c_k(t)}{\partial t}U_k(r) = \frac{1}{i\hbar}H\,\Psi(r)$$

$$= \frac{1}{i\hbar}\sum_k \big(Hc_k(t)\big)U_k(r) + \frac{1}{i\hbar}\sum_k c_k(t)\big(HU_k(r)\big)$$

it can be written that:

$$\int U_m^*(r) \sum_k \frac{\partial c_k(t)}{\partial t} U_k(r)\, d^3x$$
$$= \frac{1}{i\hbar} \int U_m^*(r) \sum_k \left(Hc_k(t)\right) U_k(r)\, d^3x$$
$$+ \frac{1}{i\hbar} \int U_m^*(r) \sum_k c_k(t)\left(HU_k(r)\right) d^3x$$

and hence:

$$\sum_k \frac{\partial c_k(t)}{\partial t} \int U_m^*(r) U_k(r) d^3x = \frac{1}{i\hbar} \sum_k \left(Hc_k(t)\right) \int U_m^*(r) U_k(r) d^3x + \frac{1}{i\hbar} \sum_k c_k(t) \int U_m^*(r)\left(HU_k(r)\right) d^3x$$

which leads to:

$$\frac{\partial c_m(t)}{\partial t} = \frac{1}{i\hbar}\left(Hc_m(t)\right) + \frac{1}{i\hbar} \sum_k c_k(t)\, H_{mk}$$

Therefore:

$$\frac{\partial c_n(t)}{\partial t} = \frac{1}{i\hbar}\left(Hc_n(t)\right) + \frac{1}{i\hbar} \sum_k c_k(t)\, H_{nk}$$

and:

$$\frac{\partial c_m^*(t)}{\partial t} = \frac{-1}{i\hbar}\left(c_m^*(t) H\right) + \frac{-1}{i\hbar} \sum_k H_{km} c_k^*(t)$$

such that finally:

$$\frac{\partial \rho_{nm}}{\partial t} = \frac{-1}{i\hbar}\left(c_m^*(t) H\right) c_n(t) + \frac{-1}{i\hbar} \sum_k H_{km} c_k^*(t)\, c_n(t) + c_m^*(t) \frac{1}{i\hbar}\left(Hc_n(t)\right) + c_m^*(t) \frac{1}{i\hbar} \sum_k c_k(t)\, H_{nk}$$

$$\frac{\partial \rho_{nm}}{\partial t} = \frac{-1}{i\hbar} \sum_k H_{km} c_k^*(t)\, c_n(t) + \frac{1}{i\hbar} \sum_k c_m^*(t) c_k(t)\, H_{nk}$$

$$\frac{\partial \rho_{nm}}{\partial t} = \frac{-1}{i\hbar} \sum_k H_{km} \rho_{nk} + \frac{1}{i\hbar} \sum_k \rho_{km} H_{nk}$$

$$\frac{\partial \rho_{nm}}{\partial t} = \frac{1}{i\hbar}\sum_{k} \rho_{km}H_{nk} - \rho_{nk}H_{km} = \frac{1}{i\hbar}[\rho, H]_{nm}$$

Each diagonal element ρ_{nn} of the density matrix operator is related to the population of a state of the physical system considered. The temporal evolution of ρ_{nn} describes the time evolution of the population of the state considered. Let N be the number of elementary systems constituting the physical system under study and τ_n the lifetime of the state $|n>$ of a system.

2) What is the origin of τ_n?

Solution:

The origin of τ_n, which is the lifetime of level n, is due to the vacuum quantum fluctuation that is responsible for the spontaneous emission.

3) Give the expression of N_n the number of systems in the state $|n>$ as a function of ρ_{nn} and N. Derive the master equation of ρ_{nn}^0 after excitation by a pulsed laser, in the absence of stimulated emission (a state of equilibrium of the level $|n>$ is assumed to be given by ρ_{nn}^0).

Solution:

From the expression $\rho_{nn} = \frac{N_n}{N}$ after excitation by a pulsed laser, the level n receives a quantity of atoms from the initial level of the transition, which leads to ρ_{nn}. Afterwards, this level labeled n will depopulate at the rate of the spontaneous emission to reach its state of equilibrium corresponding to ρ_{nn}^0. The master equation is therefore given by $\frac{\partial \rho_{nn}}{\partial t} = -\frac{\rho_{nn}}{\tau_n} + \frac{\rho_{nn}^0}{\tau_n}$.

4) Taking into account the evolution of ρ_{nn} given by the above equation, give the complete differential equation verified by ρ_{nn}.

Solution: $\frac{\partial \rho_{nn}}{\partial t} + \frac{\rho_{nn}}{\tau_n} = \frac{\rho_{nn}^0}{\tau_n} + \frac{1}{i\hbar}\sum_k \rho_{kn}H_{nk} - \rho_{nk}H_{kn}$

5) By analogy with the equation verified by ρ_{nn}, a relaxation time T_{nm} is included to describe the temporal evolution of ρ_{nn}. Give the complete differential equation verified by ρ_{nm}.

Solution:

In the equation, the population of level n from excited levels m of higher energy, whose population decreases with time with a given probability (branching ratio Φ_{nm}), is considered; then, another source term must be included on the right-hand side of the equation to take these terms into account. The terms ρ_{nn} correspond to the stimulated emissions and to the induced absorptions to which are added the depopulation terms, which result from elastic collisions and from phase shifts. The types of relaxation termed T_1 and T_2 are grouped together under the term T_{nm}.

$$\frac{\partial \rho_{nm}}{\partial t} + \frac{\rho_{nm}}{T_{nm}} = +\frac{1}{i\hbar}\sum_{k} \rho_{km}H_{nk} - \rho_{nk}H_{km}$$

2.3.4. *Application to a two-level system*

As a two-level system, a set of atoms interacting with $E(t)$, the oscillating electric field of an electromagnetic wave is considered. The angular frequency ω (or frequency) of the electric field is assumed to be close to that corresponding to the transition between the two levels $|1>$ and $|2>$ of an atom. The physical system being restricted to these two levels, the density matrix is composed of four terms.

Let H_0 be the Hamiltonian of an electron moving around the nucleus of an atom, and let H_1 be the perturbation due to the interaction of the electric field of the light wave with the electron.

The expression of H_1 is given by $-\mu E(t)$, where $\boldsymbol{\mu}$ is the dipole moment, induced by the electric field of the wave, represented by the odd operator μ. The wave is assumed to be polarized along Ox.

1) Taking into account the evolution of ρ_{nm} in the absence of relaxation, give the differential equations verified by the different components of the density matrix.

Solution:

In the absence of relaxation: $\frac{\partial \rho_{nm}}{\partial t} = +\frac{1}{i\hbar}\sum_k \rho_{km}H_{nk} - \rho_{nk}H_{km}$. Upon expansion, the following expressions are determined:

$$\frac{\partial \rho_{11}}{\partial t} = +\frac{1}{i\hbar}(\rho_{11}H_{11} - \rho_{11}H_{11}) + \frac{1}{i\hbar}(\rho_{21}H_{12} - \rho_{12}H_{21})$$

$$\frac{\partial \rho_{11}}{\partial t} = \frac{1}{i\hbar}(\rho_{21}H_{12} - \rho_{12}H_{21})$$

$$\frac{\partial \rho_{22}}{\partial t} = +\frac{1}{i\hbar}(\rho_{12}H_{21} - \rho_{21}H_{12}) + \frac{1}{i\hbar}(\rho_{22}H_{22} - \rho_{22}H_{22})$$

$$\frac{\partial \rho_{22}}{\partial t} = \frac{1}{i\hbar}(\rho_{12}H_{21} - \rho_{21}H_{12})$$

$$\frac{\partial \rho_{12}}{\partial t} = +\frac{1}{i\hbar}(\rho_{12}H_{11} - \rho_{11}H_{12}) + \frac{1}{i\hbar}(\rho_{22}H_{12} - \rho_{12}H_{22})$$

$$\frac{\partial \rho_{21}}{\partial t} = +\frac{1}{i\hbar}(\rho_{11}H_{21} - \rho_{21}H_{11}) + \frac{1}{i\hbar}(\rho_{21}H_{22} - \rho_{22}H_{21})$$

2) Explain the different terms using the properties of the hermitic operators H_0 and H_1.

Solution:

From:

$$H_{11} = \int U_1^*(H_0 + H_1)\, U_1 d^3x = \int U_1^*(H_0)\, U_1 d^3x = E_1 \int U_1^*\, U_1 d^3x = E_1$$

$$H_{22} = \int U_2^*(H_0 + H_1)\, U_2 d^3x = \int U_2^*(H_0)\, U_2 d^3x = E_2 \int U_2^*\, U_2 d^3x = E_2$$

with $\int U_1^*(H_1)\, U_1 d^3x = 0 \; since\; H_1\; is\; odd$

$$H_{12} = \int U_1^*(H_0 + H_1)\, U_2 d^3x = \int U_1^*(H_1)\, U_2 d^3x = -eE(t) \int U_1^*\, xU_2 d^3x$$

with $\int U_1^*(H_0)\,U_2 d^3x = E_2 \int U_1^*\,U_2 d^3x = 0$.

From: $\mu_{12} = \int U_1^*\,exU_2 d^3x$, the induced dipole moment,

then, $H_{12} = -\mu_{12}E(t)$.

Likewise:

$$H_{21} = \int U_2^*(H_o + H_1)\,U_1 d^3x = \int U_2^*(H_1)\,U_1 d^3x = -eE(t)\int U_2^*\,xU_1 d^3x$$

with $\int U_2^*(H_0)\,U_1 d^3x = E_1 \int U_2^*\,U_1 d^3x = 0$.

Then: $\mu_{21} = \int U_2^*\,exU_1 d^3x$, where $H_{21} = -\mu_{21}E(t) = -\mu_{12}E(t)$ because μ_{12} is real.

Thus:

$$\frac{\partial\rho_{11}}{\partial t} = \frac{\mu_{21}E(t)}{i\hbar}(\rho_{21} - \rho_{12}) = -\frac{i\mu_{21}E(t)}{\hbar}(\rho_{21} - \rho_{21}^*)$$

$$\frac{\partial\rho_{22}}{\partial t} = -\frac{i\mu_{21}E(t)}{\hbar}(\rho_{21}^* - \rho_{21})$$

$$\frac{\partial\rho_{12}}{\partial t} = -i\frac{\mu_{12}E(t)}{\hbar}(\rho_{22} - \rho_{11}) - i\rho_{12}\omega_{21}$$

$$\frac{\partial\rho_{21}}{\partial t} = +\frac{i\mu_{12}E(t)}{\hbar}(\rho_{11} - \rho_{22}) - i\rho_{21}\omega_{21}$$

3) Interpret the equations verified by ρ_{11} and ρ_{22} and show that they are physically valid.

$$\frac{\partial\rho_{11}}{\partial t} = -\frac{i\mu_{21}E(t)}{\hbar}(\rho_{21} - \rho_{21}^*)$$

$$\frac{\partial\rho_{22}}{\partial t} = -\frac{i\mu_{21}E(t)}{\hbar}(\rho_{21}^* - \rho_{21})$$

From: $\rho_{21} = a + ib$

$\rho_{21}^* = a - ib$ such that $\rho_{21} - \rho_{21}^* = 2ib$ is purely imaginary.

Hence: $\frac{\partial \rho_{11}}{\partial t} = -\frac{i\mu_{21}E(t)}{\hbar}(\rho_{21} - \rho_{21}^*) = +\frac{i\mu_{21}E(t)}{\hbar}(2ib) = \frac{2b\mu_{21}E(t)}{\hbar}$

and $\frac{\partial \rho_{22}}{\partial t} = -\frac{i\mu_{21}E(t)}{\hbar}(\rho_{21}^* - \rho_{21}) = -\frac{i\mu_{21}E(t)}{\hbar}(-2ib) = -\frac{2b\mu_{21}E(t)}{\hbar}$.

Both expressions are real, which is physically acceptable given that $\rho_{11} = N_1/N$ and $\rho_{22} = N_2/N$ are real.

4) In the dynamic regime, the populations of the states $|1>$ and $|2>$ are modified by excitation by a laser and by relaxation. Considering a pumping rate $r_n = \rho^\circ_{nn}/\tau_n$, where τ_n is the characteristic relaxation time of the level $|n>$, show that:

$$\frac{\partial(\rho_{22} - \rho_{11})}{\partial t} + \frac{\left((\rho_{22} - \rho_{11}) - (\rho_{22}^0 - \rho_{11}^0)\right)}{\tau} = -\frac{i2\mu_{21}E(t)}{\hbar}(\rho_{21} - \rho_{21}^*)$$

where τ is a characteristic time that is to be defined.

Solution:

By adding the terms corresponding to the pumping rate (the population of the level increases) and to the relaxation of the pumped level (the population of the level decreases), it can be written that:

$$\frac{\partial \rho_{11}}{\partial t} = -\frac{\rho_{11}}{\tau_1} + \frac{\rho_{11}^0}{\tau_1} + \frac{i\mu_{21}E(t)}{\hbar}(\rho_{21} - \rho_{21}^*)$$

$$\frac{\partial \rho_{22}}{\partial t} = -\frac{\rho_{22}}{\tau_1} + \frac{\rho_{22}^0}{\tau_1} - \frac{i\mu_{21}E(t)}{\hbar}(\rho_{21} - \rho_{21}^*)$$

Taking the difference:

$$\frac{\partial(\rho_{22} - \rho_{11})}{\partial t} + \frac{\left((\rho_{22} - \rho_{11}) - (\rho_{22}^0 - \rho_{11}^0)\right)}{\tau} = -\frac{i2\mu_{21}E(t)}{\hbar}(\rho_{21} - \rho_{21}^*)$$

The term $\Delta N^0 = (\rho_{22}^0 - \rho_{11}^0)$ corresponds to the population inversion of the levels following laser pumping in the absence of the field $E(t)$.

The overall relaxation time corresponds to $\frac{1}{\tau} = \frac{1}{\tau_1} + \frac{1}{\tau_2}$.

5) Similarly, by defining and interpreting a characteristic time T_2, determine the equation verified by ρ_{21}.

Solution:

A time T_2 is introduced as the time constant associated with the relaxation of the levels as a result of collisions, which causes an interruption of the phase of the wave. In the absence of a field $E(t)$, these collisions will destroy the induced macroscopic polarization $P = N\mu_{21}(\rho_{21} - \rho_{21}^*)$ by introducing a decoherence process of the phase of the atoms contributing to the macroscopic polarization.

In this case, the evolution equation of ρ_{21} is obtained as:

$$\frac{\partial \rho_{21}}{\partial t} = -\left(\frac{1}{T_2} + i\omega_{21}\right)\rho_{21} - i\frac{\mu_{21}E(t)}{\hbar}(\rho_{22} - \rho_{11})$$

In general, T_2 is interpreted as the average time between two collisions of active atoms with the atoms of the thermal bath formed by the surrounding environment.

6) Calculate $\rho_{21}(t)$ after the effect of the laser pulse (the field of the exciting wave is assumed to be zero).

Solution:

In the absence of a field $E(t)$, the differential equation is solved with the right-hand side set to zero. The solution is then given by:

$$\frac{\partial \rho_{21}}{\partial t} + \left(\frac{1}{T_2} + i\omega_{21}\right)\rho_{21} = 0$$

$$\rho_{21}(t) = \rho_{21}(0)e^{-\left(\frac{1}{T_2}+i\omega_{21}\right)t}$$

7) It is assumed that the field is oriented along Ox. Derive the polarization $P_x(t)$ of the medium (the polarization is given by $P = N\mu_{21}(\rho_{21}+\rho_{21}{}^*)$). Interpret the characteristic times.

Solution:

From:

$$\rho_{21}(t) = \rho_{21}(0)e^{-\left(\frac{1}{T_2}+i\omega_{21}\right)t}$$

and

$$\rho_{21}^*(t) = \rho_{21}^*(0)e^{-\left(\frac{1}{T_2}-i\omega_{21}\right)t}$$

it can be written that:

$$P_x(t) = N\mu_{21x}\left(\rho_{21}(0)e^{-i\omega_{21}t} + \rho_{21}^*(0)e^{+i\omega_{21}t}\right)e^{-\frac{t}{T_2}}.$$

The time T_2 is longer than the oscillation period associated with ω_{21}; there are therefore two time scales to consider in the evolution of the macroscopic polarization. Time T_2 gives the temporal envelope of the evolution of $P_x(t)$.

8) At time t, the population difference ΔN between the two levels $|1>$ and $|2>$ is given by $\Delta N = N(\rho_{22} - \rho_{11})$. Show that if ΔN° is the population difference under the equilibrium reached by laser pumping, ΔN satisfies the following equation:

$$\frac{\partial(\Delta N - \Delta N^0)}{\partial t} + \frac{(\Delta N - \Delta N^0)}{\tau} = \frac{2E_x(t)}{\hbar\omega_{21}}\left(\frac{\partial P_x(t)}{\partial t} + \frac{P_x(t)}{T_2}\right)$$

Solution:

From:

$$\frac{\partial\rho_{21}}{\partial t} = -\left(\frac{1}{T_2} + i\omega_{21}\right)\rho_{21} - i\frac{\mu_{21x}E_x(t)}{\hbar}(\rho_{22} - \rho_{11})$$

and

$$\frac{\partial\rho_{21}^*}{\partial t} = -\left(\frac{1}{T_2} - i\omega_{21}\right)\rho_{21}^* + i\frac{\mu_{21x}E_x(t)}{\hbar}(\rho_{22} - \rho_{11})$$

then:

$$\frac{\partial(\rho_{21}+\rho_{21}^*)}{\partial t}=-\frac{(\rho_{21}+\rho_{21}^*)}{T_2}-i\omega_{21}(\rho_{21}-\rho_{21}^*)$$

From:

$$\frac{\partial(\rho_{22}-\rho_{11})}{\partial t}+\frac{\left((\rho_{22}-\rho_{11})-(\rho_{22}^0-\rho_{11}^0)\right)}{\tau}=-\frac{i2\mu_{21x}E_x(t)}{\hbar}(\rho_{21}-\rho_{21}^*)$$

with the substitution $(\rho_{22}-\rho_{11})=\frac{\Delta N}{N}$ and $(\rho_{22}^0-\rho_{11}^0)=\frac{\Delta N^0}{N}$

$$(\rho_{21}-\rho_{21}^*)=\frac{i}{\omega_{21}}\left(\frac{\partial(\rho_{21}+\rho_{21}^*)}{\partial t}+\frac{(\rho_{21}+\rho_{21}^*)}{T_2}\right)$$

and

$$\left(\rho_{21}(t)+\rho_{21}^*(t)\right)=\frac{P_x(t)}{N\mu_{21x}}$$

the following expression is derived:

$$\frac{\partial\left(\frac{\Delta N}{N}\right)}{\partial t}+\frac{\left(\left(\frac{\Delta N}{N}\right)-\left(\frac{\Delta N^0}{N}\right)\right)}{\tau}=-\frac{i2\mu_{21x}E_x(t)}{\hbar}\frac{i}{\omega_{21}}\frac{1}{N\mu_{21x}}\left(\frac{\partial(P_x(t))}{\partial t}+\frac{(P_x(t))}{T_2}\right)$$

Since: $\frac{\partial\left(\frac{\Delta N^0}{N}\right)}{\partial t}=0$

it can be added to the first term on the right-hand side of the equation and thus arrive at Q.E.D:

$$\frac{\partial(\Delta N-\Delta N^0)}{\partial t}+\frac{(\Delta N-\Delta N^0)}{\tau}=\frac{2E_x(t)}{\hbar\omega_{21}}\left(\frac{\partial P_x(t)}{\partial t}+\frac{P_x(t)}{T_2}\right)$$

9) Interpret the terms appearing on the right-hand side and on the left-hand side of this equation. Under what condition does amplification take place?

Solution:

This equation shows that the energy is either in the population inversion (right-hand-side term) or in the electromagnetic field of the emitted or absorbed wave (left-hand-side term) in the light–matter interaction.

The term $(\Delta N - \Delta N^0)$ represents the difference between the inversion density at time t and at time $t = 0$ (in the absence of an emitted field $E(t)$, which polarizes the medium). The population level inversion is carried out with a pump laser at time $t = 0$, then the field $E(t)$ is spontaneously emitted as a result of vacuum fluctuations and very quickly builds up in a coherent manner (amplified stimulated emission), which means that the term $(\Delta N - \Delta N^0)$ becomes negative in this process.

This process therefore implies that the right-hand-side term averaged over a few periods of the emitted wave must be negative for the amplification to take place, namely that: $\langle \frac{2E_x(t)}{\hbar\omega_{21}}\left(\frac{\partial P_x(t)}{\partial t} + \frac{P_x(t)}{T_2}\right)\rangle < 0$.

2.4. Ising's phenomenological model for cooperative effects

The so-called Ising model (1920) was introduced by E. Ising, a student of H. Lentz, to interpret the appearance of spontaneous magnetization when a ferromagnetic body is cooled below a critical temperature called the Curie temperature. This phenomenon is due to a phase transition that takes place when a singularity is present in the free energy of the system under consideration or one of its derivatives. The most frequent examples are the transitions from the liquid phase to the gas phase, from the conductive state to the superconducting state or from the paramagnetic state to the ferromagnetic state.

The Ising model has been declined in one dimension, in two dimensions and then in larger dimensions. No transition takes place with the 1D model, and in 1944, L. Onsager gives an exact solution of the 2D model for which a phase transition is possible. This model is widely used in statistical physics in various approaches and can be used to study compounds characterized by spins at the nanoscale, spin transition compounds (or "Spin Crossover Compounds" – SCO) or thermal effects that induce cracks in alloys with the so-called ABV model.

2.4.1. *The Ising 1D model*

It is possible to solve the Ising 1D model exactly by limiting interactions to those between nearest neighbors only. *N* quantum spins are set on a 1D chain (i.e. the magnetic moments associated with each atom; spin $s = 1/2$, which can be in the state $s_z = +1/2$ or $s_z = -1/2$) take the values +1 or -1 when immersed in a magnetic field.

1) σ_i represents the component of the magnetic moment of spin i in the magnetic field *B* direction, B_0 is given by: $B_0 = \frac{g\mu_B}{2} B$, where g is the Landé factor, which is equal to 2, μ_B is the Bohr magneton, which is equal to $\mu_B = \frac{e\hbar}{2m_e}$, *e* is the charge of the electron, m_e is its mass, *h* is the Planck constant and *J* represents the coupling constant corresponding to the interaction energy between close neighbors. Give the expression of the Hamiltonian of such an interacting spin system.

Solution:

$$H_0 = -J \sum_{\langle i,j \rangle}^{N} \sigma_i \sigma_j - B_0 \sum_{i=1}^{N} \sigma_i \qquad [2.10]$$

where $\langle i, j \rangle$ denotes a summation over a pair of close neighbors, σ_i denotes the spin variable and $B_0 = \frac{g\mu_B}{2} B$ denotes the interaction between the magnetic moment and the magnetic field B.

2) Interpret each term of the equation giving the expression of the Hamiltonian. The component of the magnetic moment of spin *I* is denoted by σ_i.

Solution:

In the first term, *J* is the exchange energy, which is responsible for the cooperative behavior and the possibility of a possible phase transition. When *J* is positive, it promotes the parallel coupling of the spins (ferromagnetism), and when *J* is negative, it promotes the antiparallel coupling of the spins (anti-ferromagnetism).

In the second term, H reflects the interaction between the spins and the magnetic field present. It is the paramagnetic term that is present even though $J = 0$. This term reflects the alignment of the spins in the B field and does not participate in the cooperative effects and therefore in the phase transition.

3) Give the expression of the density operator ρ and deduce the partition function Z by giving its expression without calculating it.

Solution:

The density operator ρ is equal to: $\rho = Cexp(-\beta H)$

The normalization condition $Tr\rho = 1$ implies that: $1 = CTr\left(exp(-\beta H)\right)$

So: $C = \frac{1}{Tr\left(exp(-\beta H)\right)}$

Hence: $Z = Tr\left(exp(-\beta H)\right)$

Or by expanding: $Z = Tr\left(exp(\beta J \sum_{\langle i,j\rangle}^{N} \sigma_i\sigma_j + \beta B_0 \sum_{i=1}^{N} \sigma_i)\right)$

$$Z = \sum_{\sigma_1=\pm 1}\sum_{\sigma_2=\pm 1} \dots\dots \sum_{\sigma_N=\pm 1} exp\left(\beta J \sum_i^N \sigma_i\sigma_{i+1} + \beta B_0 \sum_{i=1}^{N} \sigma_i\right)$$

$$Z = \sum_{\sigma_1=\pm 1}\sum_{\sigma_2=\pm 1} \dots\dots \sum_{\sigma_N=\pm 1} e^{\beta J\sigma_1\sigma_2+\beta B_0\sigma_1} e^{\beta J\sigma_2\sigma_3+\beta B_0\sigma_2} \dots e^{\beta J\sigma_N\sigma_1+\beta B_0\sigma_N}$$

4) As the calculation of the partition function Z is quite tedious, proceed in steps starting from the situation where the magnetic induction B is zero. Calculate Z in the absence of an induction field B.

Solution:

In this case, Z is written as:

$$Z = then\colon \sum_{\sigma_1=\pm 1}\sum_{\sigma_2=\pm 1} \dots\dots \sum_{\sigma_N=\pm 1} e^{\beta J\sigma_1\sigma_2} e^{\beta J\sigma_2\sigma_3} \dots e^{\beta J\sigma_N\sigma_1}$$

Since each σ_i is 1 or -1, $e^{\beta J \sigma_1 \sigma_2} = e^{\beta J} or\ e^{-\beta J}$

As the pairs are regrouped and s_i are set: $s_1 = \sigma_1\sigma_2, s_2 = \sigma_2\sigma_3, \ldots s_N = \sigma_N\sigma_1$, where each s_i is 1 or -1. The periodicity condition: $\sigma_N = \sigma_1$ is applied. *Z* with the new variable is written as:

$$Z = \sum_{s_1=\pm 1} \sum_{s_2=\pm 1} \cdots\cdots \sum_{s_N=\pm 1} e^{\beta J s_1} e^{\beta J s_2} \ldots e^{\beta J s_N}$$

then

$$Z = \sum_{s_1=\pm 1} e^{\beta J s_1} \sum_{s_2=\pm 1} e^{\beta J s_2} \cdots\cdots \sum_{s_N=\pm 1} e^{\beta J s_N}$$

Thus:

$$Z = \left(\sum_{s_i=\pm 1} e^{\beta J s_i} \right)^N = \left(e^{\beta J} + e^{-\beta J}\right)^N = \left(2Ch(\beta J)\right)^N$$

5) For the calculation of the partition function *Z* in the presence of induction field *B*, given that each σ_i is equal to 1 or -1, we use a two-dimensional representation for each exponential term. In this case, we set $\Phi_{ik} = exp\left(\beta J \sigma_i \sigma_k + \frac{\beta B_0}{2}(\sigma_i + \sigma_k)\right)$, where $k = i + 1$. Given the expression of Φ, show Z as a function of the matrix Φ given by: $Z = Tr\Phi^N$.

Solution:

By setting $i = 1$ for $\sigma_i = 1$, and $k = 1$ for $\sigma_k = 1$ and $= 2\ for$ $\sigma_i = -1$, and $k = 2\ for\ \sigma_k = -1$, the expression of Φ can be written as:

$\Phi_{11} = exp\left(\beta J + \frac{\beta B_0}{2}\right)$, $\Phi_{12} = exp(-\beta J)$, $\Phi_{21} = exp(-\beta J)$ and $\Phi_{22} = exp\left(\beta J - \frac{\beta B_0}{2}\right)$. As the matrices involved in the trace are identical:

$$Z = \sum_{\sigma_1=\pm 1} \sum_{\sigma_2=\pm 1} \cdots\cdots \sum_{\sigma_N=\pm 1} \Phi\Phi \ldots \Phi = \mathrm{Tr}\Phi^N$$

6) For the calculation of the partition function Z in the presence of an induction field B, the invariance properties of the trace of a matrix are used by choosing the basis of eigenvectors, which makes it possible to make the matrix Φ diagonal in order to simplify the expression of Φ^N. Give the expression of Z as a function of the eigenvalues λ_1 and λ_2 of Φ.

Solution:

If λ_1 and λ_2 are the eigenvalues of Φ, the passage to the basis of the eigenvectors of Φ leads to the following expression: $\Phi = P\Phi' P^{-1}$, where P is the transformation matrix on the basis of the eigenvectors:

$$\Phi' = \begin{pmatrix} \lambda_1 & 0 \\ 0 & \lambda_2 \end{pmatrix}$$

In this case: $\mathrm{Tr}\Phi^N = \mathrm{Tr}\Phi'^N = \lambda_1^N + \lambda_2^N$

7) Calculate the eigenvalues λ_1 and λ_2 of Φ.

Solution:

The eigenvalue equation of Φ is written as:

$$\det(\Phi - \lambda I) = \begin{vmatrix} exp\left(\beta J + \frac{\beta B_0}{2}\right) - \lambda & exp(-\beta J) \\ exp(-\beta J) & exp\left(\beta J - \frac{\beta B_0}{2}\right) - \lambda \end{vmatrix} = 0$$

Thus:

$$\left(exp\left(\beta J + \frac{\beta B_0}{2}\right) - \lambda\right)\left(exp\left(\beta J - \frac{\beta B_0}{2}\right) - \lambda\right) - exp(-2\beta J) = 0$$

After simplification:

$$\lambda^2 - 2\lambda e^{\beta J} ch\left(\frac{\beta B_0}{2}\right) + 2sh(2\beta J) = 0$$

which leads to:

$$\lambda_{1,2} = e^{\beta J} ch\left(\frac{\beta B_0}{2}\right) \pm \sqrt{e^{2\beta J} ch^2\left(\frac{\beta B_0}{2}\right) - 2sh(2\beta J)}$$

λ_1 and λ_2 are the eigenvalues of Φ; the passage to the base of the eigenvectors of Φ leads to the following expression: $\Phi = P\Phi' P^{-1}$, where P is the transformation matrix in the basis of the eigenvectors:

$$\Phi' = \begin{pmatrix} \lambda_1 & 0 \\ 0 & \lambda_2 \end{pmatrix}$$

In this case: $\mathrm{Tr}\Phi^N = \mathrm{Tr}\Phi'^N = \lambda_1^N + \lambda_2^N$

8) Determine the partition function Z as a function of only one of the eigenvalues and show that it boils down to the same expression as that of question 4 when $B = 0$.

Solution:

Since N is large, Z is written as: $Z = Tr\Phi'^N = \lambda_1^N + \lambda_2^N = \lambda_1^N\left(1 + \lambda_2^N/\lambda_1^N\right) \cong \lambda_1^N$, where λ_1 is the greater of the two eigenvalues.

In this case, Z is given by:

$$\mathrm{Z} = \left(e^{\beta J} ch\left(\frac{\beta B_0}{2}\right) + \sqrt{e^{2\beta J} ch^2\left(\frac{\beta B_0}{2}\right) - 2sh(2\beta J)} \right)^N$$

When B = 0:

$$\mathrm{Z} = \left(e^{\beta J} + \sqrt{e^{2\beta J} - 2sh(2\beta J)} \right)^N$$

Since $2sh(2\beta J) = e^{2\beta J} - e^{-2\beta J}$,

$$e^{\beta J} + \sqrt{e^{2\beta J} - 2sh(2\beta J)} = e^{\beta J} + e^{-\beta J} = 2ch\beta J,$$

Thus:

$$Z(B=0) = (2ch\beta J)^N$$

9) By definition, the magnetic moment is given by: $\langle\mu\rangle = \sum_n \sum_m \rho_{nm}\mu_{mn} = \mu_B Tr\rho\sigma$. Deduce the expression of $\langle\mu\rangle$ as a function of Z.

Solution:

Starting from $\rho = \frac{exp\left(\beta J \sum_{\langle i,j\rangle}^{N} \sigma_i\sigma_j + \beta B_0 \sum_{i=1}^{N} \sigma_i\right)}{Z}$, $\langle\mu\rangle$ can be written as:

$$\langle\mu\rangle = \frac{\mu_B}{Z} Tr\rho\sigma = \frac{\mu_B}{Z}\ Tr\left(\sigma exp\left(\beta J \sum_{\langle i,j\rangle}^{N} \sigma_i\sigma_j + \beta B_0 \sum_{i=1}^{N} \sigma_i\right)\right)$$

where $Z = Tr\left(exp(\beta J \sum_{\langle i,j\rangle}^{N} \sigma_i\sigma_j + \beta B_0 \sum_{i=1}^{N} \sigma_i)\right)$

$$\text{As } X = \beta B_0 = \beta \frac{g\mu_B}{2} B = \beta\mu_B B = \frac{\mu_B B}{kT}$$

Then, $Z = Tr\left(exp(\beta J \sum_{\langle i,j\rangle}^{N} \sigma_i\sigma_j + X \sum_{i=1}^{N} \sigma_i)\right)$

And: $\left(\frac{\partial Z}{\partial X}\right)_\beta = \frac{\partial}{\partial X}\left\{Tr\left(exp(\beta J \sum_{\langle i,j\rangle}^{N} \sigma_i\sigma_j + X \sum_{i=1}^{N} \sigma_i)\right)\right\}$

$$= NTr\left(\sigma_i exp(\beta J \sum_{\langle i,j\rangle}^{N} \sigma_i\sigma_j + X \sum_{i=1}^{N} \sigma_i)\right) = \frac{N\langle\mu\rangle Z}{\mu_B}$$

Thus: $\langle\mu\rangle = \frac{\mu_B}{NZ}\left(\frac{\partial Z}{\partial X}\right)_\beta = \frac{\mu_B}{N}\left(\frac{\partial lnZ}{\partial X}\right)_\beta$

10) Deduce $\langle\mu\rangle$ in the absence of a magnetic induction field.

Solution:

When the magnetic induction field is zero, $\langle\mu\rangle$ is written as:

$$\langle\mu\rangle = \frac{\mu_B}{Z} Tr\rho\sigma = \frac{\mu_B}{Z}\ Tr\left(\sigma exp\left(\beta J \sum_{\langle i,j\rangle}^{N} \sigma_i\sigma_j\right)\right)$$

or: $\langle\mu\rangle = \frac{\mu_B}{NZ}\left(\frac{\partial Z}{\partial X}\right)_\beta = \frac{\mu_B}{N}\left\{\left(\frac{\partial lnZ}{\partial X}\right)_\beta\right\}_{X=0}$

From:

$$Z = \left(e^{\beta J}ch(X) + \sqrt{e^{2\beta J}ch^2(X) - 2sh(2\beta J)}\right)^N$$

$$lnZ = N\left\{ln\left(e^{\beta J}ch(X) + \sqrt{e^{2\beta J}ch^2(X) - 2sh(2\beta J)}\right)\right\}_{X=0}$$

$$\left(\frac{\partial lnZ}{\partial X}\right)_\beta = \frac{N}{Z^{1/N}}\left(e^{\beta J}sh(X) + \frac{2e^{2\beta J}ch(X)sh(X)}{2\sqrt{e^{2\beta J}ch^2(X) - 2sh(2\beta J)}}\right)$$

When $X \to 0$, $\left(\frac{\partial lnZ}{\partial X}\right)_\beta \to 0$ since $\langle\mu\rangle \propto sh(X)$; therefore, $\langle\mu\rangle = 0$.

Since $X = \beta B_0 = \beta\frac{g\mu_B}{2}B = \beta\mu_B B = \frac{\mu_B B}{kT}$, this means that for B = 0, there is no magnetic moment. The 1D crystal is paramagnetic.

In reality, the mutual interaction of the spins has an effect on its paramagnetic property. In the absence of a field, the crystal is not ferromagnetic, a property which results from the tendency of the spins to align themselves in parallel, thus minimizing the energy of the spin system. This results in a long distance order in the crystal. In a situation where the spins are aligned on a chain, each spin being coupled to its two close neighbors, it suffices for one spin not to be aligned like the others to break the order at great distance, which is more difficult to achieve in a 2D crystal.

11) Determine the internal energy *U* of the crystal as a function of *Z* in the absence of a *B* field.

Solution:

By definition, *U* is written as:

$$U = \langle H\rangle = Tr(\rho H) = \frac{1}{Z}Tr\left(H\, e^{-\beta H}\right) = -\left(\frac{\partial lnZ}{\partial\beta}\right)_{B=0}$$

Since $(B = 0) = \left(2\, ch(\beta J)\right)^N et \ln(Z) = Nln\left(2\, ch(\beta J)\right)$,

Thus: $U = -NJ\frac{2sh(\beta J)}{2ch(\beta J)} = -NJth(\beta J)$

12) Calculate U for $T = 0$ K and give an interpretation of the result obtained.

Solution:

When $T = 0$ K, $\beta \to \infty$ *and* $U = -NJ$

The spins are all parallel. As soon as T is greater than 0 K, the spin alignment is destroyed. There is no transition.

13) A chain of N ½ spins, which is characterized by an interaction between nearest neighbors, is subjected to a weak magnetic induction environment. Calculate the magnetization M and the susceptibility χ of this chain.

Solution:

From questions 9 and 10, M can be written as:

$$M = N\langle\mu_i\rangle = N\mu_B\left(\frac{\partial lnZ}{\partial X}\right)_\beta, \text{ with } X = \beta\mu_B B$$

and:

$$\left(\frac{\partial lnZ}{\partial X}\right)_\beta = \frac{N}{Z^{1/N}}\left(e^{\beta J}sh(X) + \frac{e^{2\beta J}ch(X)sh(X)}{2\sqrt{e^{2\beta J}ch^2(X) - 2sh(2\beta J)}}\right)$$

Since $\beta\mu_B B \ll 1$, $ShX = X$ and $ChX = 1$, this leads to:

$$\left(\frac{\partial lnZ}{\partial X}\right)_\beta = \frac{N}{Z^{1/N}}\left(e^{\beta J}X + \frac{e^{2\beta J}X}{\sqrt{e^{-2\beta J}}}\right) = \frac{N}{Z^{1/N}}Xe^{\beta J}\left(1 + e^{2\beta J}\right)$$
$$= \frac{N}{Z^{1/N}}Xe^{2\beta J}(2ch\beta J)$$

with $Z^{1/N} = \left(e^{\beta J}ch(X) + \sqrt{e^{2\beta J}ch^2(X) - 2sh(2\beta J)}\right) = 2ch\,\beta J$

hence: $M = N\langle\mu_i\rangle = N\mu_B\,Xe^{2\beta J}$.

The susceptibility χ of the chain is given by: $\chi = \frac{\mu_0 M}{B} = \frac{\mu_0 N\mu_B^2}{kT}e^{2J/kT}$.

14) Study the variation of the susceptibility χ at high temperatures for $J > 0$ and $J < 0$.

Solution:

For $J > 0$, the spins tend to be parallel to minimize energy.

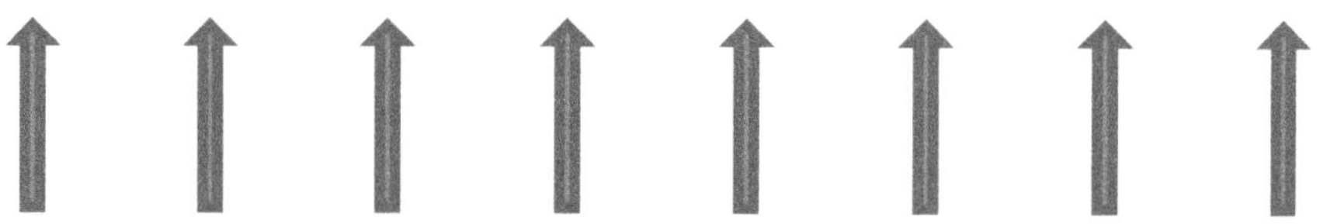

Figure 2.1. *Parallel spins. For a color version of this figure, see www.iste.co.uk/dahoo/metrology2.zip*

A $T = 0$ K, the susceptibility χ tends to infinity and any infinitesimal field B can orient the spin directions.

For large T, the susceptibility χ is written as:

$$\chi = \frac{\mu_0 N\mu_B^2}{kT}e^{2J/kT} = \frac{\mu_0 N\mu_B^2}{kT}(1 + 2J/kT) = \frac{\mu_0 N\mu_B^2}{kT}\left(1 + \frac{T_C}{T}\right) \text{ where } T_C = 2J/k$$

χ can be expressed in the form: $\chi = \frac{\mu_0 N\mu_B^2}{kT}\frac{1}{1-\frac{T_C}{T}}$.

Above T_C, the chain has a susceptibility, which is that of a ferromagnetic material.

For $J < 0$, the spins tend to be antiparallel.

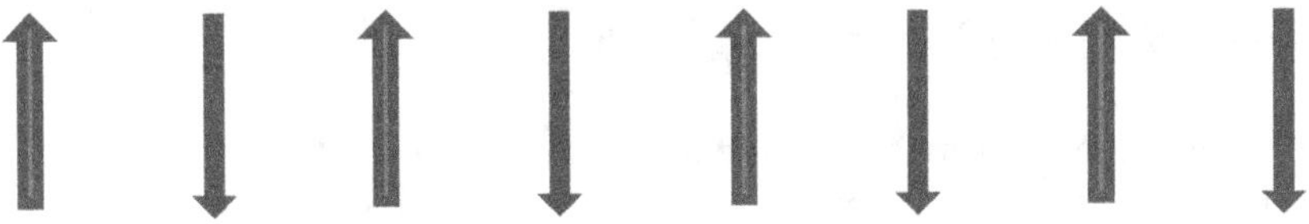

Figure 2.2. *Antiparallel spins. For a color version of this figure, see www.iste.co.uk/dahoo/metrology2.zip*

As χ is equal to zero at T = 0 K, and for large T, χ is expressed in the form: $\chi = \frac{\mu_0 N \mu_B^2}{kT}\frac{1}{1+\frac{T_C}{T}}$. Above T_C, where $T_C = 2|J|/k$, the chain has a susceptibility which is that of an antiferromagnetic material.

The 1D Ising model does not exhibit a phase transition. Fluctuations cross the chain regardless of its length.

It can be shown that for large but finite N, $ln\ Z$ and all the thermodynamic functions are analytical functions of the variable $z = e^{-2\beta\mu_B B}$. Since a phase transition is characterized by a non-analytical behavior of thermodynamic functions, the transition can only take place within the limit where N tends to infinity. Yang and Lee's theorem shows that $Z = \lim_{N\to\infty}(Z_N)^{1/N} = 0$ if $|z| = 1$ and that the thermodynamic functions are analytical functions if z is not on the circle $|z| = 1$. So, this means that the phase transition is only possible for B = 0 (z = 1). On the other hand, the phase transition is possible as soon as the dimension is greater than 1. For $D = 2$, $T_C = 2.27$ J.

If the correlation length ξ is determined from the correlation function $\langle\langle\sigma_i\sigma_j\rangle = \frac{1}{Z(B)}\sum_{\sigma_1=\pm1}\sum_{\sigma_2=\pm1}\cdots\cdots\sum_{\sigma_N=\pm1}\sigma_i\sigma_j e^{-\beta H} = e^{-r_{ij}/\xi}$, it can be shown that $\xi = \frac{1}{|lntanhJ/kT|}$. ξ is a decreasing function of temperature and tends to zero as T tends to infinity, or tends to infinity as T tends to zero. There is a competition between energy and entropy. Energy tends to align (increase order) the spins, and entropy tends to misalign (increase disorder) the spins.

The partition function acts as a generator function for mean values and correlation functions. A connected correlation function can be defined, which takes into account only the fluctuations compared to the mean values in the form: $G_{ij} = \langle\sigma_i\sigma_j\rangle_C = \langle\sigma_i\sigma_j\rangle - \langle\sigma_i\rangle\langle\sigma_j\rangle$.

From: $H = -J\sum_{\langle i,j\rangle}^{N}\sigma_i\sigma_j - \mu_B\sum_{i=1}^{N}B_i\sigma_i$

in the presence of a magnetic induction field B_i by site i and for a given dimension D, the magnetization and the connected correlation functions can be calculated.

Using:

$$Z(B) = \sum e^{-\beta H}$$

The magnetization on site i is written as:

$$M_i = \langle\sigma_i\rangle = \frac{1}{\beta\mu_B}\frac{\partial lnZ}{\partial B_i}$$

and

$$\langle\sigma_i\sigma_j\rangle = \frac{1}{(\beta\mu_B)^2 Z}\frac{\partial^2 Z}{\partial B_i\partial B_j}$$

Consequently:

$$G_{ij} = \langle\sigma_i\sigma_j\rangle_C = \langle\sigma_i\sigma_j\rangle - \langle\sigma_i\rangle\langle\sigma_j\rangle = \frac{1}{(\beta\mu_B)^2}\frac{\partial^2 lnZ}{\partial B_i\partial B_j}$$

The partition function is the generator function for the correlation functions and the free energy. This partition function is expressed as a function of lnZ. These functions are obtained by differentiation up to a given order p.

3

Quantum Optics and Quantum Computers

3.1. Introduction

Quantum computing is based on the properties of quantum mechanics. To process information, quantum computers use two-level quantum systems known as qubits (qubit means quantum bit) [NIE 10, KOK 10]. The energy states of qubits are quantized, and can be superimposed or entangled. The superposition principle of the quantum states of qubits makes it possible to perform a large number of calculations in parallel. Entanglement is used in powerful algorithms for researching massive databases, or in cryptology for analyzing encryption and detecting intrusions into information exchanges.

Quantum computers, which are currently under development, use qubits manufactured using various technologies on the nanometric scale such as superconductors [ARU 19], trapped ions, electron spins (quantum dots) [LOS 98], Rydberg atoms and Majorana fermions [KIT 03]. The calculations based on quantum algorithms are processed by unitary transformations on the states of the qubits. These transformations are carried out by quantum gates. These gates prepare the states of the qubits and make it possible to characterize them.

Whatever the qubit technology employed, the correct functioning of a quantum computer requires that the states of the qubits be controlled and that the qubits maintain their quantum properties for a period of time greater than the period of operation of the gates. The loss of coherence, which is often caused by interactions with the environment, is a phenomenon that must be analyzed and taken into account.

The objective of this chapter is to provide theoretical background and examples of applications, in order to help understand the operation of a quantum gate. The modeling of polarized light in quantum mechanics and the representation by the Bloch sphere of the states of a two-level quantum system are reviewed. The operating principle of a quantum computer is introduced. Examples of applications show how to use the Bloch sphere, predict the evolution of an initial state of a two-level quantum system and obtain the Rabi population oscillations by a coupling mechanism. Another application studies the coupling of an atom with light radiation and the effect on the Rabi oscillations of a mismatch between the frequency of the atom and the frequency of the radiation. A final exercise uses Ramsey fringes and their application to the operation of a quantum gate.

3.2. Polarized light in quantum mechanics

Studies on polarized light reveal that there are two states of polarization that correspond to a two-level energy quantum system. The study of a system in quantum mechanics is often reduced to a two-energy level system (Figure 3.1), in the same way as the two states ½ or -½ of the spin of an electron that pass from a ground state to an excited state or vice versa when a magnetic field is applied.

Figure 3.1. *Energy diagram of a two-level quantum system: a ground state | g>, an excited level | e> and a difference in energy levels $\hbar\omega$. For a color version of this figure, see www.iste.co.uk/dahoo/metrology2.zip*

When a static magnetic field is applied, the spin has a precession movement around the direction of the field. An upward precession corresponds to a spin equal to $\frac{\hbar}{2}$ and a downward precession corresponds to a spin equal to $-\frac{\hbar}{2}$. These states correspond to the eigenvalues of the spin operator $\hat{s}_z$.

A useful representation of the dynamics of the quantum states of the ½ spin of an electron is a sphere of unit radius, called the Bloch sphere (Figure 3.2).

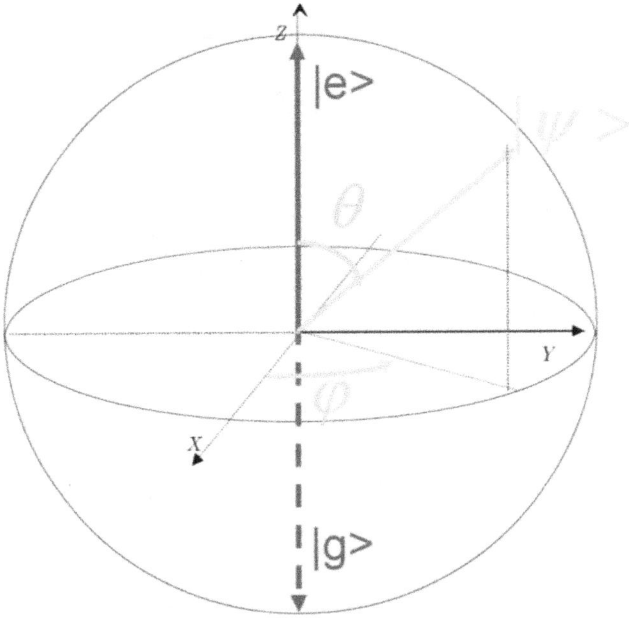

Figure 3.2. *Bloch sphere of unit radius defined by a polar angle θ and an azimuthal angle ϕ, which represent the eigenstates of a two-energy level system. For a color version of this figure, see www.iste.co.uk/dahoo/metrology2.zip*

The antipodes of the Bloch sphere correspond to the two eigenstates of the system: spin ½ or the excited state $|e\rangle$ and spin -½ or the reference ground state | $|g\rangle$.

The coordinates of a point on the sphere defined by the polar angle θ and the azimuthal angle ϕ are given by:

$$x = \sin\theta\cos\phi$$
$$y = \sin\theta\sin\phi$$
$$z = r\cos\theta$$
$$0 \leq \theta \leq \pi \ and \ 0 \leq \phi \leq 2\pi$$

In quantum mechanics, as recalled in [DAH 16], the states of a system are represented by kets $|\Psi\rangle$ in a Hilbert space and the probability P of finding the system in a state $|\Psi\rangle$ is expressed by:

$$P = |\langle\Psi|\Psi\rangle|^2$$

If $|g\rangle$ represents the spin state $-\frac{1}{2}$ and $|e\rangle$ the spin state $\frac{1}{2}$, the eigenstates are represented in a 2D space as follows:

$$|e\rangle = \begin{pmatrix} 1 \\ 0 \end{pmatrix} \quad |g\rangle = \begin{pmatrix} 0 \\ 1 \end{pmatrix}$$

The relations

$$\langle e|e\rangle = \langle g|g\rangle = 1$$
$$\langle e|g\rangle = \langle g|e\rangle = 0$$

results from the orthogonality of the states $|e\rangle$ and $|g\rangle$.

The projection operators $|e\rangle\langle e| = \begin{pmatrix} 1 & 0 \\ 0 & 0 \end{pmatrix}$ and $|g\rangle\langle g| = \begin{pmatrix} 0 & 0 \\ 0 & 1 \end{pmatrix}$ lead to the operator: $|e\rangle\langle e| - |g\rangle\langle g| = \begin{pmatrix} 1 & 0 \\ 0 & -1 \end{pmatrix}$.

Given the Pauli matrix $\sigma_z = \begin{pmatrix} 1 & 0 \\ 0 & -1 \end{pmatrix}$, this operator can be written as:

$$|e\rangle\langle e| - |g\rangle\langle g| = \sigma_z$$

The Hamiltonian operator H_0 of the $-\frac{1}{2}$ spin system in the base $(|e\rangle, |g\rangle)$ is written as:

$H_0 = \frac{\hbar\omega_0}{2}\begin{pmatrix} 1 & 0 \\ 0 & -1 \end{pmatrix} = \frac{\hbar\omega_0}{2}\sigma_z$, where $\hbar\omega_0$ is the energy difference between the two levels.

The Bloch sphere can be used to represent the space of the energy states of a two-level quantum system. The excited state $|e\rangle$ is the intersection of the sphere of unit radius with the positive z axis and the ground state $|g\rangle$ is the intersection with the negative z axis. A system state $|\Psi\rangle$ is a point on the surface of the Bloch sphere defined by the relations:

$$|\Psi\rangle = \alpha|e\rangle + \beta|g\rangle \; avec \, |\alpha|^2 + |\beta|^2 = 1$$

By setting *a priori* $\alpha = \cos\theta$ and $\beta = \sin\theta$ up to a phase factor $e^{i\phi}$, with θ varying from 0 to π and ϕ varying from 0 to 2π, then for $\theta = 0$, $|\Psi\rangle = |e\rangle$ and for $\theta = \pi$, $|\Psi\rangle = -|e\rangle$.

In this case, the ground state $|\Psi\rangle = -|g\rangle$ is not reached.

However, by setting $\alpha = \cos\frac{\theta}{2}$ and $\beta = \sin\frac{\theta}{2}$, for $\theta = 0$, $|\Psi\rangle = |e\rangle$ and for $\theta = \pi$, $|\Psi\rangle = |g\rangle$. By analogy with nuclear magnetic resonance, this corresponds to the two spin states $\frac{1}{2}$ of an electron when the magnetic field is oriented along the Oz axis.

3.3. Introduction to quantum computers

Quantum computers use two-state quantum systems as the unit for storing information. The most general quantum state, called qubit, is a state of a 2D Hilbert space that is represented by a point on the surface of the Bloch sphere:

$$|\Psi\rangle = \cos\left(\frac{\theta}{2}\right)|e\rangle + \sin\left(\frac{\theta}{2}\right)e^{i\Phi}|g\rangle .$$

An equivalent form of the qubit is:

$$|\Psi\rangle = \cos\left(\frac{\theta}{2}\right)e^{-i\Phi/2}|e\rangle + \sin\left(\frac{\theta}{2}\right)e^{i\Phi/2}|g\rangle$$

If $\theta = \frac{\pi}{2}$ and $\phi = 0$, then a qubit oriented along Ox is obtained as: $|\Psi\rangle = \frac{1}{\sqrt{2}}|e\rangle + \frac{1}{\sqrt{2}}|g\rangle$.

If $\theta = \frac{\pi}{2}$ and $\phi = \frac{\pi}{2}$, then a qubit oriented along Oy is obtained as: $|\Psi\rangle = \frac{1}{\sqrt{2}}|e\rangle + \frac{i}{\sqrt{2}}|g\rangle$.

By convention, the reference state $|g\rangle$ (the state of a spin particle $\frac{1}{2}$ whose spin is parallel to the applied static magnetic field) corresponds to the state $|1\rangle$, and the excited state $|e\rangle$ (the state of a spin particle $-\frac{1}{2}$ whose spin is antiparallel to the static field) corresponds to the state $|0\rangle$.

A qubit can be in-state $|0\rangle$ or in-state $|1\rangle$, or in any linear combination of these two states.

A quantum computer is composed of qubits, quantum registers and quantum gates.

A quantum register is a physical system that comprises more than one qubit. The state of a register is a vector of a multidimensional Hilbert space that is defined by the tensor product (denoted $\otimes$) of the states of its qubits. For a two-qubit system $|q_1\rangle$ and $|q_2\rangle$, the register wave function $|q_R\rangle$ is described by a 4D Hilbert space: $|q_R\rangle = |q_1\rangle \otimes |q_2\rangle$:

$$\text{If } |q_1\rangle = \begin{pmatrix} a \\ b \end{pmatrix} \text{ and } |q_2\rangle = \begin{pmatrix} c \\ d \end{pmatrix} \text{ then } |q_R\rangle = \begin{pmatrix} a \\ b \end{pmatrix} \otimes \begin{pmatrix} c \\ d \end{pmatrix} = \begin{pmatrix} ac \\ ad \\ bc \\ bd \end{pmatrix}$$

This tensor product can also be written by the relation:

$$|q_R\rangle = (a|0\rangle + b|1\rangle) \otimes (c|0\rangle + d|1\rangle) = ac|00\rangle + ad|01\rangle + bc|10\rangle + bd|11\rangle$$

By convention, $|00\rangle$ corresponds to the state $|0\rangle \otimes |0\rangle$ and so on.

The normalization condition is written as:

$$\left|(ac)^2\right| + \left|(ad)^2\right| + \left|(bc)^2\right| + \left|(bd)^2\right| = 1$$

A register of n qubits, corresponds to 2^n superimposed states.

A register of 10 qubits, corresponds to 1,024 superimposed states.

Consider the states of a two-qubit system described by:

$$|q_R\rangle = ac|00\rangle + ad|01\rangle + bc|10\rangle + bd|11\rangle$$

The states $|q_R\rangle$ of the qubits are either separable or entangled.

Separable states correspond to system states whose physical realities are separated. The states are separable if and only if they can be factored into the base $(|00\rangle, |01\rangle, |10\rangle, |11\rangle)$. In a quantum computer, separable states are prepared in the absence of interactions.

The entangled states cannot be factorized in the base $|(a|0\rangle + b|1\rangle) \otimes (c|0\rangle + d|1\rangle)$.

One technique for preparing entangled qubits is to bring together the degrees of freedom of the pseudo-spins qubits so that they interact with each other. When two qubits are entangled, the directions of the degrees of freedom of the spins $\frac{1}{2}$ of the pair of qubits are correlated. Although the two qubits of this pair are far apart, the correlation of the degrees of freedom is preserved.

Entanglement is a specific feature of quantum theory. It opens up many applications for data encryption and quantum teleportation. In a quantum system composed of n qubits, it is possible to obtain $2n$ separable states or $2n$ entangled states.

Quantum gates are unitary Hilbert space operators that act on isolated qubits or the qubits of quantum registers.

The action of a quantum gate P on an isolated qubit is described by the relation:

$$P|q_I\rangle = |q_F\rangle$$

where $|q_I\rangle$ is the initial state of the qubit and $|q_F\rangle$ is the final state after the gate action.

The gates of a quantum computer break down the stages of a calculation into reversible logical operations. Applying a quantum gate to its own result helps reconstruct what was applied to the gate entrance. As in a quantum computer, computing operations are reversible, and it can be shown that there is no loss of information.

When applied to a single qubit, quantum gates are analogous to rotations in the Bloch sphere. State vectors can be rotated around any axis and through any angle of rotation. Thus, any point on the Bloch sphere can be reached. The operator $R_{\hat{n}}$ corresponding to a rotation ϕ about an axis $\hat{n}$ is given by:

$$R_{\hat{n}} = \cos\left(\frac{\phi}{2}\right) I + i(\hat{n}.\sigma)\sin\left(\frac{\phi}{2}\right)$$

where I is the identity (2x2) matrix and σ is the vector of the Pauli matrices.

For example, R_Z^{ϕ}, which is the matrix of the operator representing the rotation of an angle ϕ about the Oz axis, is written as:

$$R_Z^{\phi} = e^{-i\frac{\phi}{2}} = \cos\left(\frac{\phi}{2}\right) I - i(\vec{z}.\sigma\ )\sin\left(\frac{\phi}{2}\right) = \cos\left(\frac{\phi}{2}\right)\begin{pmatrix} 1 & 0 \\ 0 & 1 \end{pmatrix} - i\sigma_z \sin\left(\frac{\phi}{2}\right)$$

$$R_Z^{\phi} = \cos\left(\frac{\phi}{2}\right)\begin{pmatrix} 1 & 0 \\ 0 & 1 \end{pmatrix} - i\begin{pmatrix} 1 & 0 \\ 0 & -1 \end{pmatrix}\sin\left(\frac{\phi}{2}\right)$$

$$R_Z^{\phi} = \begin{pmatrix} \cos\left(\frac{\phi}{2}\right) - i\sin\left(\frac{\phi}{2}\right) & 0 \\ 0 & \cos\left(\frac{\phi}{2}\right) + i\sin\left(\frac{\phi}{2}\right) \end{pmatrix} = \begin{pmatrix} e^{-i\frac{\phi}{2}} & 0 \\ 0 & e^{i\frac{\phi}{2}} \end{pmatrix}$$

The operator R_Z^{ϕ} is a rotation of an angle ϕ about the Oz axis.

The operator corresponding to a rotation $\phi = \pi$ about the Oz axis is written as:

$$R_Z^{\pi} = -i\begin{pmatrix} 1 & 0 \\ 0 & -1 \end{pmatrix}$$

The (2x2) matrix R_Y^ϕ of the operator representing the rotation of an angle ϕ about the Oy axis is given by:

$$R_Y^\phi = e^{-i\frac{\phi}{2}} = \cos\left(\frac{\phi}{2}\right)\begin{pmatrix} 1 & 0 \\ 0 & 1 \end{pmatrix} - i\begin{pmatrix} 0 & -i \\ i & 0 \end{pmatrix}\sin\left(\frac{\phi}{2}\right)$$

$$R_Y^\phi = e^{-i\frac{\phi}{2}} = \cos\left(\frac{\phi}{2}\right)\begin{pmatrix} 1 & 0 \\ 0 & 1 \end{pmatrix} - i\begin{pmatrix} 0 & -i \\ i & 0 \end{pmatrix} c = \begin{pmatrix} \cos\left(\frac{\phi}{2}\right) & -\sin\left(\frac{\phi}{2}\right) \\ \sin\left(\frac{\phi}{2}\right) & \cos\left(\frac{\phi}{2}\right) \end{pmatrix}$$

For a rotation $\phi = \pi$ about the Oy axis:

$$R_X^\pi = \begin{pmatrix} 0 & -1 \\ 1 & 0 \end{pmatrix} = -i\begin{pmatrix} 0 & -i \\ i & 0 \end{pmatrix}$$

The (2x2) matrix of the R_X^ϕ operator representing the rotation of an angle ϕ around the Ox axis is given by:

$$R_X^\phi = e^{-i\frac{\phi}{2}} = \cos\left(\frac{\phi}{2}\right)I - i(\vec{x}.\sigma)\sin\left(\frac{\phi}{2}\right) = \cos\left(\frac{\phi}{2}\right)\begin{pmatrix} 1 & 0 \\ 0 & 1 \end{pmatrix} - i\sigma_X \sin\left(\frac{\phi}{2}\right)$$

$$R_X^\phi = e^{-i\frac{\phi}{2}} = \cos\left(\frac{\phi}{2}\right)\begin{pmatrix} 1 & 0 \\ 0 & 1 \end{pmatrix} - i\begin{pmatrix} 0 & 1 \\ 1 & 0 \end{pmatrix}\sin\left(\frac{\phi}{2}\right) = \begin{pmatrix} \cos\left(\frac{\phi}{2}\right) & -i\sin\left(\frac{\phi}{2}\right) \\ -i\sin\left(\frac{\phi}{2}\right) & \cos\left(\frac{\phi}{2}\right) \end{pmatrix}$$

For a rotation $\phi = \pi$ about the Ox axis:

$$R_X^\pi = -i\begin{pmatrix} 0 & 1 \\ 1 & 0 \end{pmatrix}$$

The main quantum gates that act on an isolated qubit are the Z, X, Y Pauli gates, the $\sqrt{NOT}$ gate, the phase change gate, the S phase gate, the identity gate I and the Hadamard gate H.

– The Z Pauli gate is defined by the (2x2) matrix:

$$Z = |0\rangle\langle 0| - |1\rangle\langle 1| = \begin{pmatrix} 1 \\ 0 \end{pmatrix}(1 \quad 0) - \begin{pmatrix} 0 \\ 1 \end{pmatrix}(0 \quad 1) = \begin{pmatrix} 1 & 0 \\ 0 & -1 \end{pmatrix}$$

The action of the Z gate on the states $|+\rangle = \frac{(|0\rangle + |1\rangle)}{\sqrt{2}}$ and $|-\rangle = \frac{(|0\rangle - |1\rangle)}{\sqrt{2}}$ is given by:

$$Z|+\rangle = \frac{1}{\sqrt{2}}\begin{pmatrix} 1 & 0 \\ 0 & -1 \end{pmatrix}\begin{pmatrix} 1 \\ 1 \end{pmatrix} = \frac{1}{\sqrt{2}}\begin{pmatrix} 1 \\ -1 \end{pmatrix} = \frac{1}{\sqrt{2}}(|0\rangle - |1\rangle) = |-\rangle$$

$$Z|-\rangle = \frac{1}{\sqrt{2}}\begin{pmatrix} 1 & 0 \\ 0 & -1 \end{pmatrix}\begin{pmatrix} 1 \\ -1 \end{pmatrix} = \frac{1}{\sqrt{2}}\begin{pmatrix} 1 \\ 1 \end{pmatrix} = \frac{1}{\sqrt{2}}(|0\rangle + |1\rangle = |+\rangle$$

The Z gate reverses the $|+\rangle$ and $|-\rangle$ states. The operation of the Z gate on a qubit is equivalent to performing a rotation through an angle $\phi = \pi$ about the Oz axis.

– The Y Pauli gate:

$$Y = i|1\rangle\langle 0| - i|0\langle 1|\rangle = i\begin{pmatrix} 0 \\ 1 \end{pmatrix}(1 \quad 0) - i\begin{pmatrix} 1 \\ 0 \end{pmatrix}(0 \quad 1) = \begin{pmatrix} 0 & -i \\ i & 0 \end{pmatrix}$$

The operation of the Y gate on a qubit consists of rotating the qubit in the Bloch sphere by an angle $\phi = \pi$ about the Oy axis.

– The X gate or NOT gate:

The (2x2) matrix of the X gate is written as:

$$X = |1\rangle\langle 0| + |0\rangle\langle 1| = \begin{pmatrix} 0 \\ 1 \end{pmatrix}\begin{pmatrix} 1 & 0 \end{pmatrix} + \begin{pmatrix} 1 \\ 0 \end{pmatrix}\begin{pmatrix} 0 & 1 \end{pmatrix} = \begin{pmatrix} 0 & 1 \\ 1 & 0 \end{pmatrix}$$

The operation of the X gate on a qubit is equivalent to a rotation through an angle $n\phi = \pi$ about the Ox axis. The X gate reverses the $|0\rangle$ and $|1\rangle$ states.

Indeed:

$$X|0\rangle = \begin{pmatrix} 0 & 1 \\ 1 & 0 \end{pmatrix}\begin{pmatrix} 1 \\ 0 \end{pmatrix} = \begin{pmatrix} 0 \\ 1 \end{pmatrix} = |1\rangle$$

$$X|1\rangle = \begin{pmatrix} 0 & 1 \\ 1 & 0 \end{pmatrix}\begin{pmatrix} 0 \\ 1 \end{pmatrix} = \begin{pmatrix} 1 \\ 0 \end{pmatrix} = |0\rangle$$

– The $\sqrt{NOT}$ gate is defined by:

$$\sqrt{NOT} = \frac{1}{\sqrt{2}}\begin{pmatrix} 1 & -i \\ -i & 1 \end{pmatrix}$$

– The phase change gate R_Z^{ϕ}: rotation by an angle ϕ about the $\hat{z}$ axis

$$R_Z^{\phi} = \begin{pmatrix} 1 & 0 \\ 0 & e^{i\phi} \end{pmatrix}$$

– The *S* phase gate:

$$S = \begin{pmatrix} 1 & 0 \\ 0 & i \end{pmatrix}$$

– The identity gate I:

$$I = \begin{pmatrix} 1 & 0 \\ 0 & 1 \end{pmatrix}$$

– The Hadamard gate H corresponds to a rotation by an angle $\frac{\pi}{2}$ about the Y axis followed by a rotation by an angle π about the X axis:
$$H = \frac{1}{\sqrt{2}} \begin{pmatrix} 0 & 1 \\ 1 & 0 \end{pmatrix} \begin{pmatrix} 1 & -1 \\ 1 & 1 \end{pmatrix} = \frac{1}{\sqrt{2}} \begin{pmatrix} 1 & 1 \\ 1 & -1 \end{pmatrix}$$

This also corresponds to a rotation by an angle π about the axis $\frac{\hat{x}+\hat{z}}{\sqrt{2}}$.

The action of the H gate on a qubit $|q_1\rangle$ is summarized in Table 3.1. If the qubit is in a basis state, the H gate drives it into a superposition state in which the probability of measuring both basis states is 0.5.

As it enables quantum state superpositions, the H gate is often used in the early stages of quantum algorithms.

$\lvert q_1\rangle$	$\lvert q_0\rangle$
$\lvert 0\rangle$	$\frac{1}{\sqrt{2}}(\lvert 0\rangle+\lvert 1\rangle)$
$\lvert 1\rangle$	$\frac{1}{\sqrt{2}}(\lvert 0\rangle-\lvert 1\rangle)$
$\frac{1}{\sqrt{2}}(\lvert 0\rangle+\lvert 1\rangle)$	$\lvert 0\rangle$
$\frac{1}{\sqrt{2}}(\lvert 0\rangle-\lvert 1\rangle)$	$\lvert 1\rangle$

Table 3.1. *Action of the H gate on the states of a qubit*

The quantum gates that make it possible to perform most calculations on the registers of a quantum computer are the Hadamard gate (H), the controlled-NOT gate (*CNOT*) and the control phase gate (CΦ).

When it is applied to a two-qubit quantum register, the *CNOT* gate changes the logical state of the second qubit $|c\rangle$ (the target) if and only if the first qubit $|m\rangle$ (the control) is $|1\rangle$ (Figure 3.3). The conditional logical change of the target is a NOT inversion.

In the basis $(|00\rangle,|01\rangle,|10\rangle,|11\rangle)$, if the first qubit is the control and the second qubit is the target, the (4x4) matrix of the *CNOT* gate is:

$$CNOT = \begin{pmatrix} 1 & 0 & 0 & 0 \\ 0 & 1 & 0 & 0 \\ 0 & 0 & 0 & 1 \\ 0 & 0 & 1 & 0 \end{pmatrix}$$

Let $|m_1c_1\rangle$ and $|m_0c_0\rangle$ be the state of the quantum register before and after the application of the *CNOT* gate. The transformations carried out by the *CNOT* gate are summarized in Table 3.2.

$\|m_1c_1\rangle$	$\|m_0c_0\rangle$
$\|00\rangle$	$\|00\rangle$
$\|01\rangle$	$\|01\rangle$
$\|10\rangle$	$\|11\rangle$
$\|11\rangle$	$\|10\rangle$

Table 3.2. *Action of the CNOT gate on a two-qubit register*

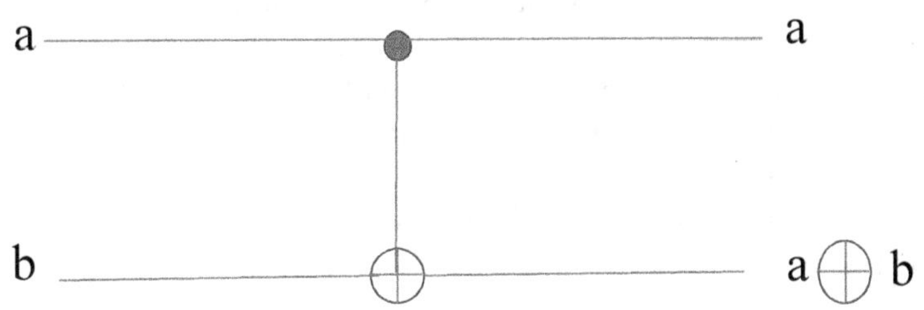

Figure 3.3. *Graphical representation of the CNOT gate acting on a two-qubit register: a is the control qubit and b is the target qubit. For a color version of this figure, see www.iste.co.uk/dahoo/metrology2.zip*

The *H* and *CNOT* gates make it possible to have two qubits interact (spin–spin interaction) and to achieve superposed and entangled states.

When two qubits are entangled, their states are correlated. For example, if the state of the first qubits is $|0\rangle$, then the state of the second qubits is $|1\rangle$.

The control phase gate CΦ applied to a two-qubit register changes the phase angle of the target qubit by an angle φ if and only if the state of the control qubit is $|1\rangle$ (Table 3.3).

$\|m_1c_1\rangle$	$\|m_0c_0\rangle$
$\|00\rangle$	$\|00\rangle$
$\|01\rangle$	$\|01\rangle$
$\|10\rangle$	$\|10\rangle$
$\|11\rangle$	$e^{i\varphi}\|11\rangle$

Table 3.3. *Action of the control phase gate CΦ on a two-qubit register*

By allowing the isolated quantum system to interact with an external system in a controlled way, the state of the quantum system under study can be measured. A quantum measurement provides probabilities for bits 1 and 0. These probabilities are positive numbers that add to 1. With a quantum measurement being able to modify the superposition of the states of the qubits, the measurements of the qubits are generally carried out at the end of the sequences of the algorithm.

A quantum measurement is described by a set of operators M_m that can operate on the state space of the system. The probability $p(m)$ that the result of a measurement is m for a state $|\psi\rangle$ is given by:

$$p(m) = \langle\psi| M_m^{\dagger} M_m |\psi\rangle$$

where $M_m^{\dagger}$ is the adjoint matrix of the operator matrix M_m.

The state $|\psi'\rangle$ of the system after the measurement is obtained by applying the operator to the current state divided by the probability of the state:

$$|\psi'\rangle = \frac{M_m |\psi\rangle}{\sqrt{p(m)}},$$

The sum of the probabilities of the results of the measurement is equal to 1:

$$\sum_m p(m) = 1 = \sum_m \langle\psi| M_m^{\dagger} M_m |\psi\rangle$$

If the measurement of the components of a superposition of qubits in the basis. $|0\rangle, |1\rangle$ is considered, then the qubit wave function can be written as:

$$|\psi\rangle = a|0\rangle + b|1\rangle$$

$$|e\rangle = |0\rangle = \begin{pmatrix} 1 \\ 0 \end{pmatrix} \quad |g\rangle = |1\rangle = \begin{pmatrix} 0 \\ 1 \end{pmatrix}$$

The measurement operators are given by:

$$M_0 = |0\rangle\langle 0| = \begin{pmatrix} 1 & 0 \\ 0 & 0 \end{pmatrix}$$

$$M_1 = |1\rangle\langle 1| = \begin{pmatrix} 0 & 0 \\ 0 & 1 \end{pmatrix}$$

Thus, $M_0^\dagger = \begin{pmatrix} 1 & 0 \\ 0 & 0 \end{pmatrix}$ and $M_1^\dagger = \begin{pmatrix} 0 & 0 \\ 0 & 1 \end{pmatrix}$

$$p(0) = \langle\psi|M_0^\dagger M_0|\psi\rangle = \begin{pmatrix} a^* & b^* \end{pmatrix}\begin{pmatrix} 1 & 0 \\ 0 & 0 \end{pmatrix}\begin{pmatrix} 1 & 0 \\ 0 & 0 \end{pmatrix}\begin{pmatrix} a \\ b \end{pmatrix}\langle 0|0\rangle = a^* a\langle 0|0\rangle = |a|^2$$

$$p(1) = \langle\psi|M_1^\dagger M_1|\psi\rangle = \begin{pmatrix} a^* & b^* \end{pmatrix}\begin{pmatrix} 0 & 0 \\ 0 & 1 \end{pmatrix}\begin{pmatrix} 0 & 0 \\ 0 & 1 \end{pmatrix}\begin{pmatrix} a \\ b \end{pmatrix}\langle 0|0\rangle = b^* b\langle 1|1\rangle = |b|^2$$

The state $|\psi'\rangle$ of the system after the measurement is given by:

$$|\psi'\rangle = \frac{M_m|\psi\rangle}{\sqrt{p(m)}}$$

For M_0, $|\psi'\rangle = \dfrac{M_0|\psi\rangle}{\sqrt{p(0)}} = \dfrac{1}{\sqrt{a^2}}\begin{pmatrix} 1 & 0 \\ 0 & 0 \end{pmatrix}\begin{pmatrix} a \\ 0 \end{pmatrix} = \dfrac{a}{|a|}|0\rangle$

Therefore, after the first measurement with the operator M_0, the state of the system is: $|\psi'\rangle = |0\rangle$.

For M_1, $|\psi'\rangle = \dfrac{M_1|\psi\rangle}{\sqrt{p(1)}} = \dfrac{1}{\sqrt{b^2}}\begin{pmatrix} 0 & 0 \\ 0 & 1 \end{pmatrix}\begin{pmatrix} 0 \\ b \end{pmatrix} = \dfrac{b}{|b|}|1\rangle$.

Thus, after the measurement with the operator M_1 the state of the system is: $|\psi'\rangle = |1\rangle$.

If the measurement operator is applied again, the state $|\psi''\rangle$ of the system after the measurement is:

For M_0, $|\psi''\rangle = \frac{a}{|a|}\frac{a}{|a|}|0\rangle = |0\rangle$.

For M_1, $|\psi''\rangle = \frac{b}{|b|}\frac{b}{|b|}|1\rangle = |1\rangle$.

As the qubit state superposition is lost after the measurement, two principles apply to quantum computers: deferred measurement at the end of the quantum circuit, and implicit measurement that consists of assuming that the qubits that are not measured at the end of circuit are also measured.

If M is a (2x2) operator performing a measurement on an isolated qubit, the operator performing a measurement on a two-qubit system is:

– $M \otimes I$, the 4x4 operator representing the measurement of M on the first qubit;

– $I \otimes M$, the operator representing the measurement of M on the second qubit.

These measurement operators can be applied to the specific state $|q_I\rangle$ of a two-qubit system defined by:

$$|q_I\rangle = \frac{1}{\sqrt{2}}(|00\rangle + |11\rangle).$$

Here, $|q_I\rangle$ can be written as the product of $|q_1\rangle = a|0\rangle + b|1\rangle$ and $|q_2\rangle >= c|0\rangle + d|1\rangle$ as follows:

$$|q_I\rangle = (a|0\rangle + b|1\rangle) \otimes (c|0\rangle + d|1\rangle) = ac|00\rangle + ad|01\rangle + bc|10\rangle + bd|11\rangle$$

$$|q_I\rangle = \frac{1}{\sqrt{2}}(|00\rangle + |11\rangle) \Rightarrow \quad ac = bd = \frac{1}{\sqrt{2}} \quad and \quad ad = bc = 0$$

This is not possible because if $ac \neq 0$ and $bd \neq 0$, then $ad \neq 0$ and $bc \neq 0$.

Since the state $|q_I\rangle$ cannot be written as the product of the states of two qubits $|q_I\rangle = (a|0\rangle + b|1\rangle) \otimes (c|0\rangle + d|1\rangle)$, the state $|q_I\rangle = \frac{1}{\sqrt{2}}(|00\rangle + |11\rangle)$ is a specific state called an entangled state.

The probability $p_1(0)$ of measuring 0 on the first qubit is:

$$p_1(0) = \langle q_I|(M_0 \otimes I)^\dagger (M_0 \otimes I)|q_I\rangle$$

$$M_0 \otimes I = \begin{pmatrix} 1 & 0 \\ 0 & 0 \end{pmatrix} \otimes \begin{pmatrix} 1 & 0 \\ 0 & 1 \end{pmatrix} = \begin{pmatrix} 1 & 0 & 0 & 0 \\ 0 & 1 & 0 & 0 \\ 0 & 0 & 0 & 0 \\ 0 & 0 & 0 & 0 \end{pmatrix}$$

$$(M_0 \otimes I)^\dagger = \begin{pmatrix} 1 & 0 & 0 & 0 \\ 0 & 1 & 0 & 0 \\ 0 & 0 & 0 & 0 \\ 0 & 0 & 0 & 0 \end{pmatrix}$$

$$(M_0 \otimes I)^\dagger (M_0 \otimes I) = \begin{pmatrix} 1 & 0 & 0 & 0 \\ 0 & 1 & 0 & 0 \\ 0 & 0 & 0 & 0 \\ 0 & 0 & 0 & 0 \end{pmatrix} \begin{pmatrix} 1 & 0 & 0 & 0 \\ 0 & 1 & 0 & 0 \\ 0 & 0 & 0 & 0 \\ 0 & 0 & 0 & 0 \end{pmatrix} = \begin{pmatrix} 1 & 0 & 0 & 0 \\ 0 & 1 & 0 & 0 \\ 0 & 0 & 0 & 0 \\ 0 & 0 & 0 & 0 \end{pmatrix}$$

$$p_1(0) = \langle q_I|(M_0 \otimes I)^\dagger (M_0 \otimes I)|q_I\rangle = \frac{1}{\sqrt{2}}(1 \quad 0 \quad 0 \quad 1) \begin{pmatrix} 1 & 0 & 0 & 0 \\ 0 & 1 & 0 & 0 \\ 0 & 0 & 0 & 0 \\ 0 & 0 & 0 & 0 \end{pmatrix} \begin{pmatrix} 1 \\ 0 \\ 0 \\ 1 \end{pmatrix} \frac{1}{\sqrt{2}} = \frac{1}{2}$$

The state $\left|q_I\right\rangle$ after the measurement is:

$$\left|q_I\right\rangle = \frac{\left(M_0 \otimes I\right)\left|q_I\right\rangle}{\sqrt{p_{1(\ )}(0)}} = \frac{\sqrt{2}}{\sqrt{2}}\begin{pmatrix} 1 & 0 & 0 & 0 \\ 0 & 1 & 0 & 0 \\ 0 & 0 & 0 & 0 \\ 0 & 0 & 0 & 0 \end{pmatrix}\begin{pmatrix} 1 \\ 0 \\ 0 \\ 1 \end{pmatrix} = \begin{pmatrix} 1 \\ 0 \\ 0 \\ 0 \end{pmatrix} = \left|00\right\rangle$$

$$I \otimes M_0 = \begin{pmatrix} 1 & 0 \\ 0 & 1 \end{pmatrix} \otimes \begin{pmatrix} 1 & 0 \\ 0 & 0 \end{pmatrix} = \begin{pmatrix} 1 & 0 & 0 & 0 \\ 0 & 0 & 0 & 0 \\ 0 & 0 & 1 & 0 \\ 0 & 0 & 0 & 0 \end{pmatrix}$$

$$\left(I \otimes M_0\right)^{\dagger} = \begin{pmatrix} 1 & 0 & 0 & 0 \\ 0 & 0 & 0 & 0 \\ 0 & 0 & 1 & 0 \\ 0 & 0 & 0 & 0 \end{pmatrix}$$

$$\left(I \otimes M_0\right)^{\dagger} \otimes \left(I \otimes M\right) = \begin{pmatrix} 1 & 0 & 0 & 0 \\ 0 & 0 & 0 & 0 \\ 0 & 0 & 1 & 0 \\ 0 & 0 & 0 & 0 \end{pmatrix}\begin{pmatrix} 1 & 0 & 0 & 0 \\ 0 & 0 & 0 & 0 \\ 0 & 0 & 1 & 0 \\ 0 & 0 & 0 & 0 \end{pmatrix} = \begin{pmatrix} 1 & 0 & 0 & 0 \\ 0 & 0 & 0 & 0 \\ 0 & 0 & 1 & 0 \\ 0 & 0 & 0 & 0 \end{pmatrix}$$

The probability of measuring 0 on the second qubits is:

$$p_2(0) = \left\langle q_I^{'}\right|\left(I \otimes M_0\right)^{\dagger}\left(M_0 \otimes I\right)\left|q_I^{'}\right\rangle = \begin{pmatrix} 1 & 0 & 0 & 0 \end{pmatrix}\begin{pmatrix} 1 & 0 & 0 & 0 \\ 0 & 0 & 0 & 0 \\ 0 & 0 & 1 & 0 \\ 0 & 0 & 0 & 0 \end{pmatrix}\begin{pmatrix} 1 \\ 0 \\ 0 \\ 0 \end{pmatrix} = 1$$

The two measurements are correlated, and these two qubits are entangled.

In a quantum computer, the entanglement of two qubits can be prepared by combining the *H* gate, the *CNOT* gate or other Pauli gates.

One way to characterize the state of a qubit is to measure the projection of its state vector onto the $\hat{z}$ axis of the Bloch sphere by applying electromagnetic field pulses of suitable phase and duration.

Starting from an initial qubit state $|0\rangle$ along the $\hat{z}$ axis, it is possible to obtain:

– a specific excited qubit state rotating about the x or y axis by applying electromagnetic field pulses X_π or Y_π;

– a superposition of states by applying the pulses $X_{\frac{\pi}{2}}$ or $Y_{\frac{\pi}{2}}$.

Qubits can be physically produced by various two-level quantum systems such as the following: superconductive circuits made up of nanometer-scale superconductive inductors and capacitors, trapped ions, electrons in quantum dots, rotations of nuclear spins of molecules by static and rotating magnetic fields, atoms with circular Rydberg states and cold atoms manipulated by laser. The preparations of the qubit states carried out in the quantum gates are achieved, according to the chosen implementation, by laser beams, radio frequency waves, or electromagnetic fields of appropriate amplitudes, durations and phases.

The disturbances exerted by the surrounding environment on a quantum information storage system can introduce random phase variations and errors in a calculation. However, the measurement techniques, which have been developed in the field of nuclear magnetic resonance to filter the effects of the disturbances of the surrounding environment on the magnetic spin states [VAN 04], can be applied to qubits that are based on other physical principles, in order to prepare the states of qubits, characterize them and reduce the error rate.

To build a quantum computer, it is necessary that [DIV 97]:

– the physical system chosen for implementing the quantum bits should provide the greatest possible number of qubits;

– the states of the qubits are prepared and controlled;

– the time during which the qubits lose their initial or prepared state be much longer than the operating time of the gates. The loss of the initial state of a qubit is called decoherence. Decoherence may be due to a transition from one specific state to another (relaxation) or to a loss of the initial phase (phase shift);

– the entangled qubits are prepared;

– the transformations performed by the quantum gates have a very low error rate;

– the measurements of the states of the qubits are precise.

Figure 3.4 shows the steps of a computation in a quantum computer that has 3 qubits. The input of the computation is the initial state of the quantum register. Each calculation step corresponds to the action of quantum gates. Some gates (*H*, *X*, *Y* gates) act on isolated qubits. Others act on two qubits, similar to the *CNOT* gates. The final state of the quantum register is measured at the end of the circuit and the measurement of the state of the register is the output of the calculation.

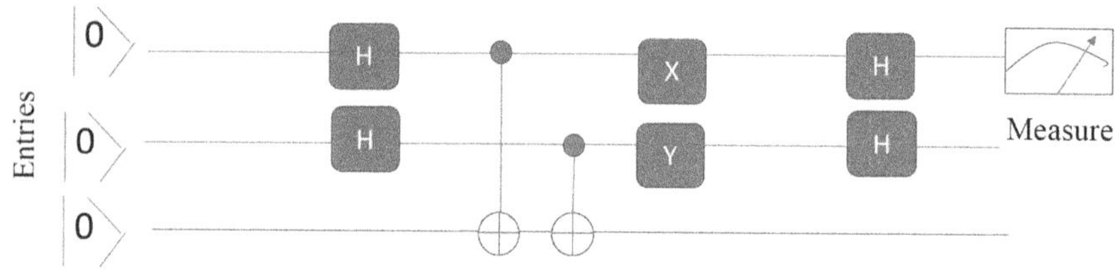

Figure 3.4. *Schematic diagram of a quantum computation. The time goes from left to right. Each horizontal line represents a qubit. The boxes represent the H, X and Y gates, and the measurement system unit. The dots represent the CNOT gates. For a color version of this figure, see www.iste.co.uk/dahoo/metrology2.zip*

The first algorithm was developed by D. Deutsch in 1985, which demonstrated that quantum computers outperform classical computers [DEU 85, DEU 92, LIN 98, CHU 98]. The objective of this algorithm is to determine with the minimum number of evaluations of this function that a binary function $f(x):\{0.1\}\rightarrow\{0.1\}$ is either constant and always returns 0 or 1, i.e. $(f(0)=f(1))$, or is balanced (i.e. delivers either 1 or 0 for 50% of cases).

When this algorithm is performed on a conventional computer, it is necessary to process three queries: calculate $f(0)$, $f(1)$, then compare the results. On a quantum computer, Deutch used state superposition to demonstrate that only one evaluation is enough.

Deutsch's algorithm is achieved by associating a two-qubit system with an *H* gate, a *U* gate and an *H* gate on the first qubit (Figure 3.5). Measuring the output state of the first qubit makes it possible to conclude whether the binary function is constant or not.

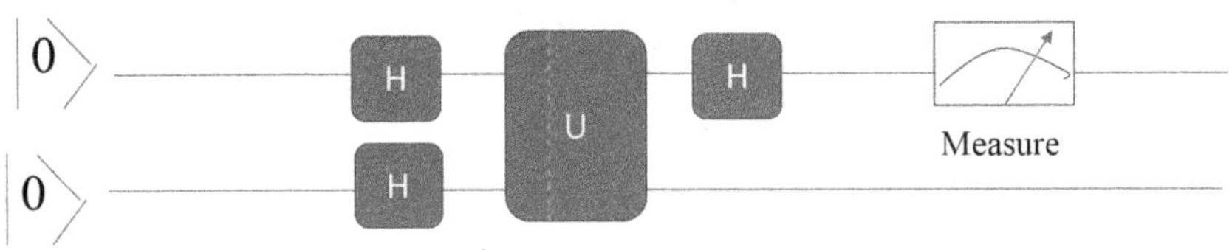

Figure 3.5. *Realization of the Deutsch algorithm in a quantum computer with a two-qubit register, H gates, a U gate and a measurement system. For a color version of this figure, see www.iste.co.uk/dahoo/metrology2.zip*

If the initial state of the two qubits is $|0\rangle|0\rangle$, the action of the *H* gate on the two qubits is:

$$(H\otimes H)|00\rangle=(H|0\rangle\otimes H|0\rangle)=\left(\frac{(|0\rangle+|1\rangle)}{\sqrt{2}}\right)\otimes\left(\frac{(|0\rangle+|1\rangle)}{\sqrt{2}}\right)$$

The *U* gate, which acts on qubits $|x\rangle$ and $|a\rangle$, performs a unitary operation that is associated with the binary function *f* under study. Its output state is given by:

$$U|x\rangle|a\rangle=(-1)^{f(x)}|x\rangle|a\rangle$$

After the action of the *H* gate, the *U* gate and the *H* gate, the output state $|\psi_1\rangle$ of the first qubit is:

$$|\psi_1\rangle=HUH|0\rangle=HU\left(\frac{|0\rangle+|1\rangle}{\sqrt{2}}\right)=H\left(\frac{1}{\sqrt{2}}\right)\left((-1)^{f(0)}|0\rangle+(-1)^{f(1)}|1\rangle\right)$$

$$|\psi_1\rangle = HUH|0\rangle = H(-1)^{f(0)}\left(\frac{1}{\sqrt{2}}\right)\left(|0\rangle + (-1)^{f(0)}(-1)^{f(1)}|1\rangle\right)$$

$$|\psi_1\rangle = HUH|0\rangle = H(-1)^{f(0)}\left(\frac{1}{\sqrt{2}}\right)\left(|0\rangle + (-1)^{f(0)\oplus f(1)}|1\rangle\right)$$

$$|\psi_1\rangle = HUH|0\rangle = (-1)^{f(0)}\left(\frac{1}{\sqrt{2}}\right)\left(|f(0)\oplus f(1)\rangle\right)$$

The output state measurement is $|\psi_1\rangle = |0\rangle$ if and only if $f(0)\oplus f(1) = 0$, i.e. if *f* is constant.

Two other algorithms are often used to justify the interest of quantum computers. These are the Grover algorithm [GRO 96], which allows very fast searches in databases, and the Shor algorithm [SHO 97], which is used in cryptanalysis.

3.4. Preparing a qubit

3.4.1. *Application of the Bloch sphere*

The effects of a polarizer can be understood as the eigenstates of a spin observable directed along the z axis.

The Hamiltonian operator of the system is written as:

$$H_0 = \frac{\hbar\omega_0}{2}\begin{pmatrix} 1 & 0 \\ 0 & -1 \end{pmatrix}$$

where $\hbar\omega_0$ is the energy between the two levels.

If $\sigma_x, \sigma_y, \sigma_z$ are the Pauli matrices, then:

$$H_0 = \frac{\hbar\omega_0}{2}\sigma_z$$

1) Demonstrate that any state of the system can be written in the form:

$$\left|\Psi_{\pm}\right\rangle = \cos\left(\frac{\theta}{2}\right)e^{-i\Phi/2}\left|e\right\rangle + \sin\left(\frac{\theta}{2}\right)e^{i\Phi/2}\left|g\right\rangle \qquad [5.3]$$

Propose a geometric interpretation of this state.

What is the probability P_e of detecting the system in the excited state?

2) Assuming that the initial state of the system is:

$$\left|\Psi(t=0)\right\rangle = \cos\left(\frac{\theta}{2}\right)\left|e\right\rangle| + \sin\left(\frac{\theta}{2}\right)\left|g\right\rangle$$

Describe the evolution of the system. Propose a geometric interpretation. This phenomenon is usually called "Larmor precession".

3) A coupling is connected between the two levels. The Hamiltonian is written in a vector basis as:

$$H = \frac{\hbar\omega_0}{2}\begin{pmatrix} 1 & 0 \\ 0 & -1 \end{pmatrix} + \frac{\hbar\Omega_0}{2}\begin{pmatrix} 0 & 1 \\ 1 & 0 \end{pmatrix}$$

What are the eigenstates of the coupled system? Describe the temporal evolution of any state of this system. Propose a geometric interpretation. How does P_e evolve? This phenomenon is usually called "Rabi precession".

3.4.1.1. *Answer to question 1*

In section 4.3.2 of [DAH 16], we express that the most general state of spin $\boldsymbol{\sigma}$ in a direction defined on the Bloch sphere by the polar angles θ and azimuthal angles φ is expressed in the Pauli matrices basis by:

$$\vec{\sigma} = \sin\theta\cos\varphi\,\sigma_x + \sin\theta\sin\varphi\,\sigma_y + \cos\theta\,\sigma_z = \begin{pmatrix} \cos\theta & e^{-i\varphi}\sin\theta \\ e^{i\varphi}\sin\theta & -\cos\theta \end{pmatrix}$$

where: $I = \begin{pmatrix} 1 & 0 \\ 0 & 1 \end{pmatrix}; \sigma_z = \begin{pmatrix} 1 & 0 \\ 0 & -1 \end{pmatrix}; \sigma_x = \begin{pmatrix} 0 & 1 \\ 1 & 0 \end{pmatrix}; \sigma_y = \begin{pmatrix} 0 & -i \\ i & 0 \end{pmatrix}$.

Writing that det $(\sigma - \lambda I) = 0$, the eigenvalue equation is obtained as:

$$-(\cos^2\theta - \lambda^2) - \sin^2\theta = 0 \Rightarrow \lambda^2 = 1$$

The eigenvalues are ± 1. The associated eigenvectors are determined from the equation:

$$(\sigma - \lambda I)|\Psi\rangle = 0$$

with $|\Psi\rangle = \begin{pmatrix} \Psi_1 \\ \Psi_2 \end{pmatrix} = \Psi_1|e\rangle + \Psi_2|g\rangle$ and $|\Psi_1|^2 + |\Psi_2|^2 = 1$.

For $\lambda = 1$:

$(\cos\theta - 1)\Psi_1 + \sin\theta e^{-i\phi}\Psi_2 = 0$, and when considering half angle, the formula can be written as:

$$-2\sin^2\frac{\theta}{2}\Psi_1 + 2\sin\frac{\theta}{2}\cos\frac{\theta}{2}e^{-i\phi}\Psi_2 = 0$$

$$\Psi_1 = \frac{\cos\frac{\theta}{2}}{\sin\frac{\theta}{2}}e^{-i\phi}\Psi_2$$

The normalization condition leads to:

$$\left(\left(\frac{\cos^2\frac{\theta}{2}}{\sin^2\frac{\theta}{2}}\right) + 1\right)|\Psi_2|^2 = 1$$

$$|\Psi_2| = \sin\frac{\theta}{2}$$

$$|\Psi_1| = \cos\frac{\theta}{2}$$

Choosing Ψ_2 $\Psi_2 = \sin\frac{\theta}{2}e^{\frac{i\phi}{2}}$, Ψ_1 can be written as: $\Psi_1 = \cos\frac{\theta}{2}e^{-\frac{i\phi}{2}}$

Therefore:

$$|\Psi_+\rangle = \cos(\frac{\theta}{2})e^{-i\Phi/2}|e\rangle + \sin(\frac{\theta}{2})e^{i\Phi/2}|g\rangle$$

$$\textit{if } \theta = 0 \textit{ and } \phi = 0 \textit{ then } |\Psi_+\rangle = |e\rangle$$
$$\textit{if } \theta = \pi \textit{ and } \phi = 0 \textit{ then } |\Psi_+\rangle = |g\rangle$$

$\phi = 0$, as this phase factor is not involved in the probability calculation. The solution $|\Psi_+\rangle = \cos(\frac{\theta}{2})e^{-i\Phi/2}|e\rangle + \sin(\frac{\theta}{2})e^{i\Phi/2}|g\rangle$ is general.

For $\lambda = -1$:

$$(\cos\theta + 1)\Psi_1 + \sin\theta e^{-i\phi}\Psi_2 = 0$$

Using the half-angle formula, this can be written as:

$$2\cos^2\frac{\theta}{2}\Psi_1 + 2\sin\frac{\theta}{2}\cos\frac{\theta}{2}e^{-i\phi}\Psi_2 = 0$$

$$\Psi_1 = -\frac{\sin\frac{\theta}{2}}{\cos\frac{\theta}{2}}e^{-i\phi}\Psi_2$$

The normalization condition leads to:

$$((\frac{\sin^2\frac{\theta}{2}}{\cos^2\frac{\theta}{2}}) + 1)|\Psi_2|^2 = 1$$

$$|\Psi_2| = \cos\ \frac{\theta}{2}$$

$$|\Psi_1| = \sin\ \frac{\theta}{2}$$

Choosing Ψ_2

$$\Psi_2 = \cos\frac{\theta}{2} e^{\frac{i\phi}{2}}, \; \Psi_1 \text{ is given by:}$$

$$\Psi_1 = -\sin\frac{\theta}{2} e^{-\frac{i\phi}{2}}$$

Furthermore: $\left|\Psi_-\right\rangle = -\sin(\frac{\theta}{2})e^{-i\Phi/2}\left|g\right\rangle + \cos(\frac{\theta}{2})e^{i\Phi/2}\left|e\right\rangle$

$$\textit{if } \theta = 0 \textit{ and } \phi = 0 \textit{ then} \left|\Psi_-\right\rangle = \left|e\right\rangle$$
$$\textit{if } \theta = \pi \textit{ and } \phi = 0 \textit{ then} \left|\Psi_+\right\rangle = -\left|g\right\rangle$$

As the eigenvectors $\left|\Psi_+\right\rangle$ and $\left|\Psi_-\right\rangle$ are parallel but in opposite directions, they are antiparallel.

Of the two solutions, only one $|\Psi_+>$ has a physical meaning because it explains the data of the observable spin model.

Geometric interpretation:

The state of the system is on a quantization axis that makes an angle θ with the Oz axis and an angle ϕ with the xoz plane. The new poles are the intersections of $\left|\Psi_+\right\rangle$ and $\left|\Psi_-\right\rangle$ with the sphere.

The probability P_e of detecting the system in the excited state is given by:

$$\left(\left\langle e\middle|\Psi_+\right\rangle\right)^2 = \sin^2\left(\frac{\theta}{2}\right)$$

3.4.1.2. *Answer to question 2*

At time $t = 0$, the system is defined by:

$$\left|\Psi(t=0)\right\rangle = \cos\left(\frac{\theta}{2}\right)\left|e\right\rangle| + \sin\left(\frac{\theta}{2}\right)\left|g\right\rangle$$

At time t, the system is defined by:

$$|\Psi(t)\rangle = a(t)|e\rangle + b(t)|g\rangle$$

The temporal evolution of the state vectors is obtained by the Schrödinger equation:

$$i\hbar\frac{\partial|\Psi(t)\rangle}{\partial t} = H|\Psi(t)\rangle$$

By replacing $|\Psi(t)\rangle$ with its expression, and then by identification, it becomes:

$$i\hbar\frac{\partial a(t)}{a(t)}|e\rangle = \frac{\hbar\omega_0}{2}\sigma_z|e\rangle = \frac{\hbar\omega_0}{2}|e\rangle$$

$$i\hbar\frac{\partial b(t)}{b(t)}|g\rangle = \frac{\hbar\omega_0}{2}\sigma_z|g\rangle = -\frac{\hbar\omega_0}{2}|g\rangle$$

$$a(t) = a(0)e^{-i\frac{\omega_0 t}{2}}$$

$$b(t) = b(0)e^{+i\frac{\omega_0 t}{2}}$$

where:

$$|\Psi(t)>= a(0)e^{-i\frac{\omega_0 t}{2}}|e> +b(0)e^{+i\frac{\omega_0 t}{2}}|g>$$

$$a(0) = \cos\frac{\theta}{2}$$

$$b(0) = \sin\frac{\theta}{2}$$

The temporal evolution of the state vector is therefore equal to:

$$|\Psi(t)\rangle = e^{-i\frac{\omega_0 t}{2}}\cos\frac{\theta}{2}|e\rangle + e^{+i\frac{\omega_0 t}{2}}\sin\frac{\theta}{2}|g\rangle$$

In the basis $(|e\rangle, |g\rangle)$, $|\Psi(t)\rangle$ is written as:

$$|\Psi(t)\rangle = \begin{pmatrix} e^{-i\frac{\omega_0 t}{2}} & 0 \\ 0 & e^{+i\frac{\omega_0 t}{2}} \end{pmatrix} \begin{pmatrix} \cos\frac{\theta}{2} \\ \sin\frac{\theta}{2} \end{pmatrix}$$

The state vector makes an angle θ with the quantization axis and rotates around the Oz axis with an angular velocity ω_0. This angular frequency is the Larmor precession frequency. On the Bloch sphere, $\omega_0 t$ corresponds to ϕ.

3.4.1.3. *Answer to question 3*

The Hamiltonian operator of the coupled system is written as:

$$\hat{H} = \frac{\hbar\omega_0}{2}\begin{pmatrix} 1 & 0 \\ 0 & -1 \end{pmatrix} + \frac{\hbar\Omega_0}{2}\begin{pmatrix} 0 & 1 \\ 1 & 0 \end{pmatrix}$$

$$\hat{H} = \frac{\hbar}{2}\begin{pmatrix} \omega_0 & \Omega_0 \\ \Omega_0 & -\omega_0 \end{pmatrix}$$

The eigenvalues are calculated from the solutions of $\det(\hat{H} - \lambda I) = 0$. Thus:

$$\det\begin{pmatrix} \frac{\hbar\omega_0}{2} - \lambda & \frac{\hbar\Omega_0}{2} \\ \frac{\hbar\Omega_0}{2} & \frac{-\hbar\omega_0}{2} - \lambda \end{pmatrix} = (\frac{\hbar\omega_0}{2} - \lambda)(\frac{-\hbar\omega_0}{2} - \lambda) - (\frac{\hbar\Omega_0}{2})^2 = 0$$

$$\Rightarrow \quad \lambda = \pm\frac{\hbar\sqrt{\omega^2{}_0 + \Omega^2{}_0}}{2}$$

Setting:

$$\Omega = \sqrt{\omega_0{}^2 + \Omega_0{}^2}$$

Then, the E_+ and E_- energies of the eigenstates are written as:

$$E_+ = \frac{\hbar}{2}\Omega$$
$$E_- = -\frac{\hbar}{2}\Omega$$

As the matrix of the coupled system $\hat{H}$ is a 2x2 matrix, the following formula can be used to calculate the two eigenvalues:

$$E_\pm = \frac{Tr\hat{H} \pm \sqrt{\left(Tr\hat{H}\right)^2 - 4\det\hat{H}}}{2}$$

$$Tr\hat{H} = 0$$

$$\det\hat{H} = -\frac{\hbar^2}{4}\left(\omega_0^2 + \Omega_0^2\right) = -\frac{\hbar^2}{4}\Omega^2$$

Thus:

$$E_+ = \frac{\hbar}{2}\Omega$$
$$E_- = -\frac{\hbar}{2}\Omega$$

The Hamiltonian operator of the coupled system can be written as:

$$\hat{H} = \frac{\hbar}{2}\begin{pmatrix} \omega_0 & \Omega_0 \\ \Omega_0 & -\omega_0 \end{pmatrix} = \frac{\hbar}{2}\begin{pmatrix} E_1 & \Omega_0 \\ \Omega_0 & E_2 \end{pmatrix}$$

It should be noted that:

$$E_1 > E_2$$

and that $\left(E_1 - E_2\right) > 0$

$$E_1 - E_+ \;= E_1 - \frac{Tr\hat{H} + \sqrt{\left(Tr\hat{H}\right)^2 - 4\det\hat{H}}}{2}$$

$$E_1 - E_+ = \frac{\hbar\omega_0}{2} + \frac{\hbar}{2}\sqrt{\left(\omega_0^2 + \Omega_0^2\right)} = \frac{\hbar\omega_0}{2} + \frac{\hbar\omega_0}{2}\sqrt{\left(1 + \frac{\Omega_0^2}{\omega_0^2}\right)}$$

Setting $\tan\theta = \frac{\Omega_0}{\omega_0}$:

$$E_1 - E_+ = \frac{\hbar\omega_0}{2}(1 + \sqrt{\left(1 + \tan^2\theta\right)})$$

$$\sqrt{\left(1 + \tan^2\theta\right)} = \frac{1}{\cos\theta}$$

Thus:

$$E_1 - E_+ = \frac{\hbar\omega_0}{2}\left(\frac{\cos\theta + 1}{\cos\theta}\right) = \frac{\hbar\omega_0}{2}\left(\frac{2\cos^2\frac{\theta}{2}}{\cos\theta}\right)$$

Likewise:

$$\left(E_2 - E_+\right) = -\frac{\hbar}{2}\left(\omega_0 + \Omega\right) = -\frac{\hbar}{2}\omega_0 + \frac{\hbar}{2}\sqrt{\left(\omega_0^2 + \Omega_0^2\right)}$$

$$\left(E_2 - E_+\right) = \frac{\hbar}{2}\left(-1 + \frac{1}{\cos\theta}\right) = \frac{\hbar\omega_0}{2}\left(\frac{2\sin^2\frac{\theta}{2}}{\cos\theta}\right)$$

Finally:

$$E_1 - E_- = -E_2 - E_- = -\left(E_2 + E_-\right) = -\left(E_2 - E_+\right) = -\frac{\hbar\omega_0}{2}\left(\frac{2\sin^2\frac{\theta}{2}}{\cos\theta}\right)$$

$$E_2 - E_- = -E_1 - E_- = -E_1 + E_+ = -\left(E_1 - E_+\right) = -\frac{\hbar\omega_0}{2}\left(\frac{2\cos^2\frac{\theta}{2}}{\cos\theta}\right)$$

The eigenvectors $|\psi_+\rangle$ and $|\psi_-\rangle$, which are written in the form:

$$|\psi_+\rangle = \cos\frac{\theta}{2}e^{-\frac{i\phi}{2}}|e\rangle + \sin\frac{\theta}{2}e^{+\frac{i\phi}{2}}|g\rangle$$
$$|\psi_-\rangle = -\sin\frac{\theta}{2}e^{+\frac{i\phi}{2}}|e\rangle + \cos\frac{\theta}{2}e^{-\frac{i\phi}{2}}|g\rangle$$

are solutions if the following equations hold:

$$\frac{\hbar}{2}\begin{pmatrix} \omega_0 + \Omega & \Omega_0 \\ \Omega_0 & -\omega_0 + \Omega \end{pmatrix}\begin{pmatrix} \cos\frac{\theta}{2}e^{-i\frac{\phi}{2}} \\ \sin\frac{\theta}{2}e^{+i\frac{\phi}{2}} \end{pmatrix} = \begin{pmatrix} 0 \\ 0 \end{pmatrix}$$

and

$$\frac{\hbar}{2}\begin{pmatrix} \omega_0 - \Omega & \Omega_0 \\ \Omega_0 & -\omega_0 - \Omega \end{pmatrix}\begin{pmatrix} \sin\frac{\theta}{2}e^{-i\frac{\phi}{2}} \\ -\cos\frac{\theta}{2}e^{+i\frac{\phi}{2}} \end{pmatrix} = \begin{pmatrix} 0 \\ 0 \end{pmatrix}$$

By setting $A = \frac{\hbar}{2}(\omega_0 - \Omega)\sin\frac{\theta}{2}e^{-i\frac{\phi}{2}} - \frac{\hbar}{2}(\Omega_0)\cos\frac{\theta}{2}e^{+i\frac{\phi}{2}}$,

the above equation holds if $A = 0$.

If $\phi = 0$, then:

$$A = (E_1 - E_+)\sin\frac{\theta}{2} - \frac{\hbar}{2}\Omega_0 \cos\frac{\theta}{2}$$

$$A = \frac{\hbar}{2}\omega_0 \frac{2\cos^2\frac{\theta}{2}}{\cos\theta}\sin\frac{\theta}{2} - \frac{\hbar}{2}\Omega_0 \cos\frac{\theta}{2} = \frac{\hbar}{2}\omega_0 \cos\frac{\theta}{2}\left(\tan\theta - \frac{\Omega_0}{\omega_0}\right) = 0$$

according to the definition of $\tan\theta$.

By setting $B = \frac{\hbar}{2}(\Omega_0)\sin\frac{\theta}{2}e^{-i\frac{\phi}{2}} + \frac{\hbar}{2}(\omega_0 + \Omega_0)\cos\frac{\theta}{2}e^{+i\frac{\phi}{2}}$,

the above equation holds if $B = 0$.

If $\phi = 0$, then:

$$B = \frac{\hbar}{2}\Omega_0 \sin\frac{\theta}{2} + \frac{\hbar}{2}(\omega_0 + \Omega_0)\cos\frac{\theta}{2} = \frac{\hbar}{2}\Omega_0 \sin\frac{\theta}{2} - (E_2 - E_+)\cos\frac{\theta}{2}$$

$$B = \frac{\hbar}{2}\Omega_0 \sin\frac{\theta}{2} - \frac{\hbar}{2}\omega_0 \frac{2\sin^2\frac{\theta}{2}}{\cos\theta}\cos\frac{\theta}{2}$$

$B = \frac{\hbar}{2}\omega_0 \sin\frac{\theta}{2}\left(\frac{\Omega_0}{\omega_0} - \tan\theta\right) = 0$, according to the definition of $\tan\theta$.

By setting $C = \frac{\hbar}{2}(\omega_0 + \Omega)\cos\frac{\theta}{2}e^{-i\frac{\phi}{2}} + \frac{\hbar}{2}(\Omega_0)\sin\frac{\theta}{2}e^{+i\frac{\phi}{2}}$, the above equation holds if $C=0$.

If $\phi=0$, then:

$$C=\frac{\hbar}{2}(\omega_0+\Omega)\cos\frac{\theta}{2}+\frac{\hbar}{2}(\Omega_0)\sin\frac{\theta}{2}=(E_1-E_-)\cos\frac{\theta}{2}+\frac{\hbar}{2}(\Omega_0)\sin\frac{\theta}{2}$$

$$C=-\frac{\hbar}{2}\omega_0\frac{2\sin^2\frac{\theta}{2}}{\cos\theta}\cos\frac{\theta}{2}+\frac{\hbar}{2}(\Omega_0)\sin\frac{\theta}{2}$$

$C=\frac{\hbar}{2}\omega_0\sin\frac{\theta}{2}\left(-\tan\theta+\frac{\Omega_0}{\omega_0}\right)=0$, according to the definition of $\tan\theta$.

By setting $D=\frac{\hbar}{2}(\Omega_0)\cos\frac{\theta}{2}e^{-i\frac{\phi}{2}}-\frac{\hbar}{2}(\omega_0-\Omega)\sin\frac{\theta}{2}e^{+i\frac{\phi}{2}}$,

the above equation holds if $D=0$.

If $\phi=0$, then:

$$D=\frac{\hbar}{2}(\Omega_0)\cos\frac{\theta}{2}-\frac{\hbar}{2}(\omega_0-\Omega)\sin\frac{\theta}{2}$$

$$D=\frac{\hbar}{2}(\Omega_0)\cos\frac{\theta}{2}-\frac{\hbar}{2}(E_1-E_+)\sin\frac{\theta}{2}$$

$$D=\frac{\hbar}{2}(\Omega_0)\cos\frac{\theta}{2}-\frac{\hbar}{2}\omega_0\frac{2\cos^2\frac{\theta}{2}}{\cos\theta}\sin\frac{\theta}{2}=\frac{\hbar}{2}(\omega_0)\cos\frac{\theta}{2}\left(\frac{\Omega_0}{\omega_0}-\tan\theta\right)=0$$

according to the definition of $\tan\theta$.

In conclusion, the eigenvectors $|\psi_+\rangle$ and $|\psi_-\rangle$, which are written in the form:

$$|\psi_+\rangle = \cos\frac{\theta}{2}e^{-\frac{i\phi}{2}}|e\rangle + \sin\frac{\theta}{2}e^{+\frac{i\phi}{2}}|g\rangle$$

$$|\psi_-\rangle = -\sin\frac{\theta}{2}e^{+\frac{i\phi}{2}}|e\rangle + \cos\frac{\theta}{2}e^{-\frac{i\phi}{2}}|g\rangle$$

are solutions of the equation $\hat{H}|\psi\rangle = \frac{\hbar}{2}\begin{pmatrix}\Omega & 0\\ 0 & -\Omega\end{pmatrix}|\psi\rangle$ with:

$$\Omega = \sqrt{\omega_0^{\ 2} + \Omega_0^{\ 2}} \text{ and } \tan\theta = \frac{\Omega_0}{\omega_0}.$$

The temporal evolution $|\psi(t)\rangle$ of any state of the system is described by the Schrödinger equation:

$$i\hbar\frac{d|\Psi(t)\rangle}{dt} = H|\Psi(t)\rangle$$

$$|\Psi(t)\rangle = b_e(t)|e\rangle + b_g(t)|g\rangle, \text{ with } |b_e|^2 + |b_g|^2 = 1$$

then:

$$i\hbar\frac{db_e}{dt} = \frac{\hbar}{2}\Omega b_g$$

$$i\hbar\frac{db_{eg}}{dt} = \frac{\hbar}{2}\Omega b_e$$

Thus:

$$\frac{d^2b_e}{dt^2} + \left(\frac{\Omega}{2}\right)^2 b_e = 0$$

$$\frac{d^2 b_g}{dt^2} + \left(\frac{\Omega}{2}\right)^2 b_g = 0$$

The solutions that satisfy these equations and the normalization condition $|b_e|^2 + |b_g|^2 = 1$ are of the form:

$$b_e = A\cos\left(\frac{\Omega t}{2}\right) + B\,s\,in\left(\frac{\Omega t}{2}\right)$$

$$b_g = i\left(B\cos\left(\frac{\Omega t}{2}\right) - A\sin\left(\frac{\Omega t}{2}\right)\right)$$

$$\frac{d^2 b_e}{dt^2} + \left(\frac{\Omega}{2}\right)^2 b_e = 0$$

$$\Rightarrow \quad -A\left(\frac{\Omega}{2}\right)^2 \cos\frac{\Omega t}{2} - B\left(\frac{\Omega}{2}\right)^2 \sin\frac{\Omega t}{2} = -\left(\frac{\Omega}{2}\right)^2 \left(A\cos\frac{\Omega t}{2} + B\sin\frac{\Omega t}{2}\right)$$

$$\Rightarrow \quad B = 0$$

$$\Rightarrow \quad A = 1$$

$$b_e = \cos\left(\frac{\Omega t}{2}\right)$$
$$(
$$
$$b_g = -i\sin\left(\frac{\Omega t}{2}\right)$$

The probability P_e of detecting the system in the excited state is:

$$\left(\langle e|\Psi(t)\rangle\right)^2 = \sin^2\left(\frac{\Omega t}{2}\right)$$

Coupling in the considered frame of reference results in the transitions between the two energy levels of the quantum system. During these transitions, the system transits from a lower energy state to a higher energy state and vice versa. These are the periods of the Rabi oscillations $T_{Rabi} = \frac{2\pi}{\Omega}$, with $\Omega = \sqrt{\omega_0^2 + \Omega_0^2}$.

3.5. Application: interaction of a qubit with a classical field

In the framework of the interaction of a two-state atom with a classical electric field within the dipolar approximation, the Hamiltonian of the atom–field coupling is written as:

$$V = -D.E$$

where D is the atomic dipole operator and E is the applied electric field: $E = E_0 \cos(\omega t - \phi)$.

On the basis $(|e\rangle, |g\rangle)$ of the previously studied atom, the operator D is written as:

$$D = d_{eg}\left(|e\rangle\langle g| + |g\rangle\langle e|\right)$$

where $d_{eg} = \langle e|qR|g\rangle$ is the dipole matrix element, taken to be real, between the two atomic levels of the energy transition.

1) Give the expression of the Hamiltonian of the system in the basis (|e>, |g>) and describe the Schrödinger equation governing the evolution of the atom. The classical field will be written as the superposition of two rotating fields in opposite directions, setting $d_{eg}.E_0 = \hbar\Omega_0$.

2) $|\Phi\rangle$ is the transform of $|\Psi\rangle$ by the unit transformation R defined by:

$$R = \begin{pmatrix} e^{\frac{i\omega t}{2}} & 0 \\ 0 & e^{-\frac{i\omega t}{2}} \end{pmatrix}$$

What is the geometric interpretation of this transformation?

Give the equation of motion for $|\Phi>$. Note that $\delta = \omega_0 - \omega$ is the detuning between the atom and the field. Demonstrate that the term coming from the interaction with the rotating field in the retrograde direction (approximation of the rotating wave) can be dropped. How do $|\Phi>$, $|\Psi>$ geometrically evolve?

What happens in the case $\delta = 0$? What is meant by giving a $\pi/2$ pulse and a π pulse to the atom?

3.5.1. *Answer to question 1*

A two-level atom system is comparable to a ½ spin. The Hamiltonian operator of the atom can be written as:

$$\tilde{H}_0 = \frac{\hbar\omega_0}{2}\left(|e\rangle\langle e| - |g\rangle\langle g|\right)$$

See equation [5.38] of [DAH 16].

In the basis $\left(|e\rangle, |g\rangle\right)$:

$$\tilde{H}_0|e\rangle = \frac{\hbar\omega_0}{2}\left(|e\rangle\langle e|e\rangle - |g\rangle\langle g|e\rangle\right) = \frac{\hbar\omega_0}{2}|e\rangle$$

$$\tilde{H}_0|g\rangle = \frac{\hbar\omega_0}{2}\left(|e\rangle\langle e|g\rangle - |g\rangle\langle g|g\rangle\right) = -\frac{\hbar\omega_0}{2}|g\rangle$$

where:

$$H_0 = \frac{\hbar\omega_0}{2}\begin{pmatrix} 1 & 0 \\ 0 & -1 \end{pmatrix} = \frac{\hbar\omega_0}{2}\sigma_z$$

The Hamiltonian of the atom–field coupling is written as:

$$\tilde{V} = -\tilde{D}.E(r,t)$$

$$D = d_{eg}\left(|e\rangle\langle g| + |g\rangle\langle e|\right)$$

$$V = (-d_{eg}\, E_0 |e\rangle\langle g| - d_{eg}\, E_0 |g\rangle\langle e|).\cos(\omega t - \phi)$$

$$\left(V|e\rangle = -d_{eg}\, E_0\left(|g\rangle\langle e|e\rangle\right).\cos(\omega t - \phi) = -d_{eg}\, E_0 |g\rangle.\cos(\omega t - \phi)\right)$$

$$V|g\rangle = -d_{eg}\, E_0\left(|e\rangle\langle g|g\rangle\right).\cos(\omega t - \phi) = -d_{eg}\, E_0 |e\rangle.\cos(\omega t - \phi)$$

or in matrix form:

$$V = -d_{eg}\, E_0.\cos(\omega t - \phi)\begin{pmatrix} 0 & 1 \\ 1 & 0 \end{pmatrix}$$

if

$$\sigma_+ = \frac{1}{2}(\sigma_x + i\sigma_y) = \frac{1}{2}\begin{pmatrix} 0 & 1 \\ 1 & 0 \end{pmatrix} + \frac{1}{2}\begin{pmatrix} 0 & 1 \\ -1 & 0 \end{pmatrix} = \begin{pmatrix} 0 & 1 \\ 0 & 0 \end{pmatrix}$$

$$\sigma_- = \frac{1}{2}(\sigma_x - i\sigma_y) = \frac{1}{2}\begin{pmatrix} 0 & 1 \\ 1 & 0 \end{pmatrix} - \frac{1}{2}\begin{pmatrix} 0 & 1 \\ -1 & 0 \end{pmatrix} = \begin{pmatrix} 0 & 0 \\ 1 & 0 \end{pmatrix}$$

then:

$$V = -d_{eg}\, E_0(\sigma_+ + \sigma_-).\cos(\omega t - \phi)$$

This corresponds to the interaction Hamiltonian of the Jaynes–Cummings model described in equation [5.39] of [DAH 16] by setting:

$$D = d_{eg}\left(|e\rangle\langle g| + |g\rangle\langle e|\right) = d_{eg}(\sigma_+ + \sigma_-)$$

and the Hamiltonian described in equation [6.24] of [DAH 16], i.e.:

$$\hat{H} = \varepsilon_1 | g >< g | + \varepsilon_2 | e >< e | + \hat{\mu} E(t)\left(| g >< e | + | e >< g |\right)$$

setting:

$$\varepsilon_1 = -\frac{\hbar\omega_0}{2}$$

$$\varepsilon_2 = +\frac{\hbar\omega_0}{2}$$

and

$$\hat{\mu} = d_{eg}$$

$$E(t) = E_0 \cos(\omega t - \phi)$$

If d_{eg} is real, then setting: $d_{eg}.E_0 = \hbar\Omega_0$

$$V = -\hbar\Omega_0(\sigma_+ + \sigma_-).\cos(\omega t - \phi)$$

The Hamiltonian of the system is then written as:

$$H = H_0 + V = \frac{\hbar\omega_0}{2}\sigma_z - \hbar\Omega_0(\sigma_+ + \sigma_-).\cos(\omega t - \phi)$$

If $|\Psi(t_0)\rangle$ is the state of the system at time t_0, assuming that this system is not subject to any observation during the interval $[t_0, t]$, the state of the system at the time t is determined by the equation:

$|\Psi(t)\rangle = U(t,t_0)|\Psi(t_0)\rangle$, where $U(t,t_0)$ is the system evolution operator.

$U(t,t_0)$ is a linear operator that satisfies $U(t_0,t_0) = 1$.

$U(t,t_0)$ is a unitary operator.

In Schrödinger's representation (see section 3.4 of Chapter 3 of [DAH 16]), the time evolution of the quantum state of the atom is given by:

$$i\hbar\frac{\partial|\Psi(t)\rangle}{\partial t}=H|\Psi(t)\rangle$$

if the system is isolated, i.e. not subjected to an observation.

Thus:

$$\frac{d}{dt}|\Psi(t)\rangle=\left(\frac{d}{dt}U(t,t_0)\right)|\Psi(t_0)\rangle \text{ and}$$

$$\frac{d}{dt}U(t,t_0)=\frac{H}{i\hbar}U(t,t_0)$$

Consequently:

$$U(t,t_0)=1-\frac{i}{\hbar}\int_{t_0}^{t}HU(t',t_0)dt'$$

3.5.2. *Answer to question 2*

R is a unitary operator. If R is applied to an eigenvector $|\Psi(t=0)\rangle=\cos\frac{\theta}{2}|e\rangle+\sin\frac{\theta}{2}|g\rangle$, then:

$$\hat{R}|\Psi\rangle=\begin{pmatrix} e^{\frac{i\omega t}{2}} & 0 \\ 0 & e^{-\frac{i\omega t}{2}} \end{pmatrix}\begin{pmatrix} \cos\frac{\theta}{2} \\ \sin\frac{\theta}{2} \end{pmatrix}=\begin{pmatrix} e^{\frac{i\omega t}{2}}\cos\frac{\theta}{2} \\ e^{-\frac{i\omega t}{2}}\sin\frac{\theta}{2} \end{pmatrix}$$

By multiplying the space vector by $e^{\frac{i\omega t}{2}}$, this equation can be written as:

$$\hat{R}\left|e^{-\frac{i\omega t}{2}}\Psi\right\rangle=\begin{pmatrix} \cos\frac{\theta}{2} \\ e^{-i\omega t}\sin\frac{\theta}{2} \end{pmatrix}$$

On the Bloch sphere, the R transformation rotates a state vector about the Oz axis at the radial speed ω. The R transformation is used to describe a frame of reference revolving around the Oz axis. The coupling electric field comprises two terms: one that oscillates in phase but at twice the angular frequency 2ω, and the other that oscillates in opposition to the radial frequency $(\omega \approx \omega_0)$. As the integral over time of the term that oscillates at twice the frequency does not provide any contribution, only a constant term remains in the rotating frame of reference. This is called the rotating wave approximation (RWA).

If at time t, in a fixed frame of reference, the system is defined by the wave function:

$$|\Psi(t)\rangle = a_e(t)|e\rangle + a_g(t)|g\rangle$$

In the rotating frame of reference, it is defined by the wave function $|\Phi(t)\rangle$:

$$|\Phi(t)\rangle = \hat{R}|\Psi(t)\rangle$$

In Schrödinger's representation, the time evolution of the quantum state of the atom is given by:

$$i\hbar\frac{d|\Phi(t)\rangle}{dt} = \hat{H}_{eff}|\Phi(t)\rangle$$, where $\hat{H}_{eff}$ is independent of time.

$$i\hbar\frac{d|\hat{R}(t)\Psi(t)\rangle}{dt} = \hat{H}_{eff}\hat{R}(t)\Psi(t)$$

$$i\hbar\frac{d|\hat{R}(t)\Psi(t)\rangle}{dt} = i\hbar\frac{d\hat{R}(t)}{dt}|\Psi(t)\rangle + \hat{R}(t)\hat{H}(t)|\Psi(t)\rangle$$

$$\hat{H}_{eff}\hat{R}(t)|\Psi(t)\rangle = i\hbar\frac{d\hat{R}(t)}{dt}|\Psi(t)\rangle + \hat{R}(t)\hat{H}(t)|\Psi(t)\rangle$$

$$\widehat{H}_{eff} = i\hbar \frac{d\widehat{R}(t)}{dt} \widehat{R}^{\dagger}(t) + \widehat{R}(t)\widehat{H}(t)\widehat{R}^{\dagger}(t)$$

$$\widehat{H}(t) = \frac{\hbar}{2}\begin{pmatrix} \omega_0 & \Omega_0 e^{-i\omega t} \\ \Omega_0 e^{+i\omega t} & -\omega_0 \end{pmatrix}$$

$$\widehat{R}(t)\widehat{H}(t)\widehat{R}^{\dagger}(t) = \frac{\hbar}{2}\widehat{R}(t)\begin{pmatrix} \omega_0 & \Omega_0 e^{-i\omega t} \\ \Omega_0 e^{+i\omega t} & -\omega_0 \end{pmatrix}\begin{pmatrix} e^{-i\omega t} & 0 \\ 0 & e^{+i\omega t} \end{pmatrix}$$

$$= \frac{\hbar}{2}\widehat{R}(t)\begin{pmatrix} \omega_0 e^{-i\omega t} & \Omega_0 \\ \Omega_0 & -\omega_0 e^{+i\omega t} \end{pmatrix}$$

$$\widehat{R}(t)\widehat{H}(t)\widehat{R}^{\dagger}(t) = \frac{\hbar}{2}\begin{pmatrix} e^{i\omega t} & 0 \\ 0 & e^{-i\omega t}{}_0 \end{pmatrix}\begin{pmatrix} \omega_0 e^{-i\omega t} & \Omega_0 \\ \Omega_0 & -\omega_0 e^{+i\omega t} \end{pmatrix} = \frac{\hbar}{2}\begin{pmatrix} \omega_0 & \Omega_0 \\ \Omega_0 & -\omega_0 \end{pmatrix}$$

$$i\hbar \frac{d\widehat{R}(t)}{dt} \widehat{R}^{\dagger}(t) = i\hbar\begin{pmatrix} i\omega e^{i\omega t} & 0 \\ 0 & -i\omega e^{-i\omega t} \end{pmatrix}\begin{pmatrix} e^{-i\omega t} & 0 \\ 0 & e^{i\omega t} \end{pmatrix} = \hbar\begin{pmatrix} -\omega & 0 \\ 0 & \omega \end{pmatrix}$$

$$\widehat{H}_{eff} = \frac{\hbar}{2}\begin{pmatrix} \omega_0 - \omega & \Omega_0 \\ \Omega_0 & -(\omega_0 - \omega) \end{pmatrix} = \frac{\hbar}{2}\begin{pmatrix} -\delta & \Omega_0 \\ \Omega_0 & \delta \end{pmatrix}$$

$$\delta = \omega - \omega_0$$

The eigenvalues are obtained by solving $\det(\hat{H}_{eff} - \lambda I) = 0$ Thus:

$$\det \frac{\hbar}{2}\begin{pmatrix} -\delta - \lambda & \Omega_0 \\ \Omega_0 & \delta - \lambda \end{pmatrix} = \frac{\hbar}{2}\left((-\delta - \lambda)(\delta - \lambda) - \Omega_0{}^2\right) = 0$$

The eigenvalues are: $\tilde{H} = -\frac{\hbar}{2}\sqrt{\delta^2 + \Omega_0^2}\begin{pmatrix} 1 & 0 \\ 0 & -1 \end{pmatrix} = -\frac{\hbar}{2}\sqrt{\delta^2 + \Omega_0^2}\,\widehat{\sigma}_z$.

If $\tan(\theta) = -\frac{\Omega_0}{\delta}$, then the eigenvectors are:

$$\left|\tilde{\Psi}_+\right\rangle = \cos\frac{\theta}{2}\left|e\right\rangle + \sin\frac{\theta}{2}\left|g\right\rangle$$

$$\left|\tilde{\Psi}_-\right\rangle = -\sin\frac{\theta}{2}\left|e\right\rangle + \cos\frac{\theta}{2}\left|g\right\rangle$$

To obtain the solution in a fixed coordinate system, the wave function is expanded in stationary states:

$$\left|\tilde{\Psi}(t)\right\rangle = \left\langle\tilde{\Psi}_-\middle|\tilde{\Psi}_0\right\rangle e^{it\frac{\sqrt{\delta^2+\Omega_0^2}}{2}}\left|\tilde{\Psi}_-\right\rangle + \left\langle\tilde{\Psi}_+\middle|\tilde{\Psi}_0\right\rangle e^{-it\frac{\sqrt{\delta^2+\Omega_0^2}}{2}}\left|\tilde{\Psi}_+\right\rangle$$

if $\left|\tilde{\Psi}_0\right\rangle = \left|g\right\rangle$, then:

$$\left|\tilde{\Psi}(t)\right\rangle = \left\langle\tilde{\Psi}_-\middle|g\right\rangle e^{it\frac{\sqrt{\delta^2+\Omega_0^2}}{2}}\left|\tilde{\Psi}_-\right\rangle + \left\langle\tilde{\Psi}_+\middle|g\right\rangle e^{-it\frac{\sqrt{\delta^2+\Omega_0^2}}{2}}\left|\tilde{\Psi}_+\right\rangle$$

$$\left|\tilde{\Psi}(t)\right\rangle = \cos\frac{\theta}{2} e^{it\frac{\sqrt{\delta^2+\Omega_0^2}}{2}}\left|\tilde{\Psi}_-\right\rangle + \sin\frac{\theta}{2} e^{-it\frac{\sqrt{\delta^2+\Omega_0^2}}{2}}\left|\tilde{\Psi}_+\right\rangle$$

$$\left|\tilde{\Psi}(t)\right\rangle = \cos\frac{\theta}{2} e^{it\frac{\sqrt{\delta^2+\Omega_0^2}}{2}}\left(-\sin\frac{\theta}{2}\left|e\right\rangle + \cos\frac{\theta}{2}\left|g\right\rangle\right) + \sin\frac{\theta}{2} e^{-it\frac{\sqrt{\delta^2+\Omega_0^2}}{2}}$$

$$\cos\frac{\theta}{2}\left|e\right\rangle + \sin\frac{\theta}{2}\left|g\right\rangle$$

The probability P_e of detecting the system in the excited state is:

$$P_e = \left|\left\langle e\middle|\tilde{\Psi}(t)\right\rangle\right|^2$$

$$\left\langle e\middle|\tilde{\Psi}(t)\right\rangle = \cos\frac{\theta}{2} e^{it\frac{\sqrt{\delta^2+\Omega_0^2}}{2}}\left(-\sin\frac{\theta}{2}\right) + \sin\frac{\theta}{2} e^{-it\frac{\sqrt{\delta^2+\Omega_0^2}}{2}}\cos\frac{\theta}{2}$$

$$P_e = \sin^2\frac{\theta}{2}\cos^2\frac{\theta}{2}\left(e^{-it\frac{\sqrt{\delta^2+\Omega_0^2}}{2}} - e^{-it\frac{\sqrt{\delta^2+\Omega_0^2}}{2}}\right)^2 = \sin^2\theta\sin^2\left(t\frac{\sqrt{\delta^2+\Omega_0^2}}{2}\right)$$

$$P_e = \frac{1}{1+\frac{\delta^2}{\Omega_0^2}}\sin^2\left(t\frac{\sqrt{\delta^2+\Omega_0^2}}{2}\right)$$

This equation means that the population of the two states of the quantum system oscillates when the atom is coupled to light radiation. The frequency and amplitude of the population oscillations of the ground and excited states depend on the amplitude of the electric field of the coupling radiation, the mismatch between the frequency of the transitions of the atom and the frequency of the coupling light radiation.

If the mismatch between the frequency of the atom transitions and the frequency of the radiation is zero $\delta = 0$, then $|\Phi(t)\rangle$ is represented by a vector in the xoy plane perpendicular to 0z.

Giving a $\frac{\pi}{2}$ pulse to the atom is like rotating through $\frac{\pi}{2}$ the angle of the equivalent spin of the atom, in the following way:

– When the spin is initially oriented along Oz, a $\frac{\pi}{2}$ pulse causes it to orient itself in the xoy plane, for example along Ox.

– When the spin is initially oriented along Ox, a $\frac{\pi}{2}$ pulse causes it to rotate about Ox in the xoz plane.

Giving a π pulse to the atom is like rotating the angle of the equivalent spin of the atom, in the following way:

– When the spin is initially oriented along Ox, a pulse π causes it to rotate by a 180° angle about Ox. The spin system is still in the xoy plane but has undergone a mirror symmetry.

– When the spin is initially oriented along Oz, a π pulse causes it to rotate by a 180° angle, for example, from the excited state to the ground state or vice versa.

The pulse sequences $\frac{\pi}{2}$ and π are used in quantum gates (*H*, *CNOT*, etc.) to prepare, transform and measure qubit states. For example, $\frac{\pi}{2}$ pulses make it possible to achieve a qubit state superposition.

3.6. Applying Ramsey fringes to evaluate the duration of phase coherence

In the case where the atom and the field are very slightly detuned, the atom that is initially in the | *g*> state experiences a first $\pi / 2$ pulse and then the interaction is quenched. The atom and the field move freely for a time T_{vol}. The interaction is then switched on again and the atom undergoes a second $\pi / 2$ pulse.

1) Give as a function of the detuning δ the evolution of the probability of measuring the atom in the state | *e*> after the two pulses.

2) Give an interpretation of this phenomenon in terms of interference fringes.

This technique, which is known as "Ramsey interferometry", is widely used in metrology and quantum optics. Give some examples of application.

3.6.1. *Answer to question 1*

In the case where the atom and the field are very slightly detuned, it can be written that $\delta \simeq 0 \; or \; \omega \simeq \omega_0$.

The initial time evolution of the state vector is given by:

$$|\Psi(t)\rangle = \left(e^{-i\frac{\omega_0 t}{2}} \cos\frac{\theta}{2} |e\rangle + \sin\frac{\theta}{2} e^{+i\frac{\omega_0 t}{2}} |g\rangle \right)$$

As the atom experiences a first pulse $\pi/2$: $\theta=\dfrac{\pi}{2}$

$$\left|\Psi(t)\right\rangle=\frac{1}{\sqrt{2}}\left(\left|e\right\rangle+ie^{+i\omega_0 t}\left|g\right\rangle\right)$$

If the relative phase is given by $\Phi=\omega_0 t=\dfrac{\pi}{2}$, then:

$$\left|\Psi(t=\frac{\pi}{2\omega_0})\right\rangle=\frac{1}{\sqrt{2}}\left(\left|e\right\rangle-\left|g\right\rangle\right)$$

The duration τ of the $\pi/2$ pulse has the effect of dividing the initial population of the spins oriented initially along | g> into two populations of the same weight, oriented according to the eigenstates of the Hamiltonian. The spins are oriented according to the plane (Ox, Oy).

The atom and the field move freely (precession movement around Ox) during the duration T_{vol} of the first pulse $\pi/2$. However, the components that will be oriented along the x axis at the end of T_{vol} will remain oriented along the x axis after applying the second pulse $\pi/2$ (the final spin of the atom is oriented along $\left|e\right\rangle$).

When there is a resonance between the precession frequency and the frequency of the rotating frame of reference, the final state does not vary with time T_{vol}.

When the atom and the field are very slightly detuned, the Bloch vector rotates in the x, y plane by an angle equal to δT_{vol}. If the atom experiences a second pulse $\pi/2$, the probability P_e of detecting the atom in the excited state after the second pulse $\pi/2$ is:

$$P_e=|\left\langle e|\Psi\right\rangle|^2=\frac{\Omega_0^2}{\delta^2+\Omega_0^2}\sin^2\left(\frac{\sqrt{(\delta^2+\Omega_0^2)}t}{2}\right)$$

The final spin is oriented according to $|e\rangle$ and $|g\rangle$. The final state is tilted about $|g\rangle$ by an angle δT_{vol}.

3.6.2. *Answer to question 2*

By varying the duration of the pulse or the frequency detuning, the probability of detecting the atom in the excited state $|e\rangle$ varies periodically. The probability of the excited state $|e\rangle$ varies as a function of the detuning in a similar way to interference fringes. These excited state population variations are called Ramsey fringes (Figure 3.6).

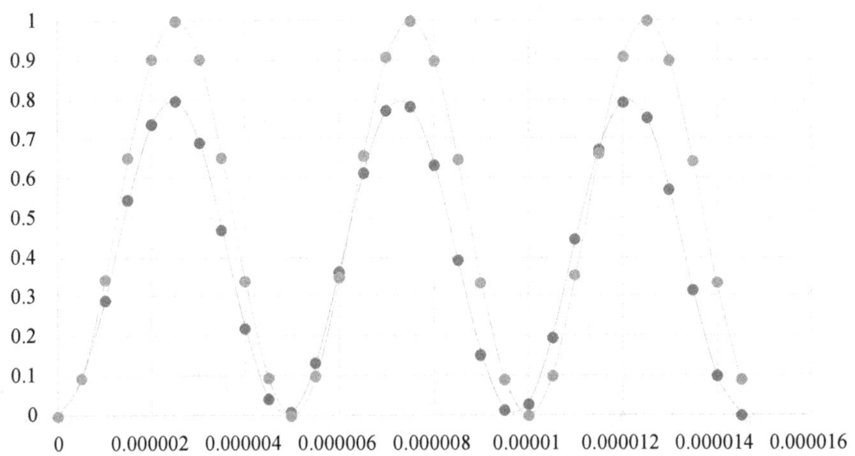

Figure 3.6. *Variation of the probability of the excited state population as a function of time (Ramsey fringes). The orange curve corresponds to the case where the detuning is zero. The blue curve has a relative detuning of 0.25. For a color version of this figure, see www.iste.co.uk/dahoo/metrology2.zip*

If the detuning increases:

– the frequency of the population oscillations increases;

– the amplitude of the population oscillations decreases.

The observed decay on the Ramsey fringes can be used to measure the frequency of the precession. The decrease in the population of the excited state characterizes the time during which the qubit maintains phase coherence. If the typical duration of a quantum logic gate is known, this duration can be compared to the time during which the qubit maintains phase coherence. This approach makes it possible to estimate the number of logical operations that can be performed for a very low error rate.

By applying two successive $\pi/2$ Ramsey pulses, the control of the states of a qubit can be achieved [LIM 14].

If the first $\pi/2$ Ramsey pulse rotates the pseudo-spin from the direction $|0\rangle$ of the z axis to the direction of the x axis. A superposition of states in the x, y plane is achieved.

If a second delayed $\pi/2$ Ramsey pulse is applied, various probabilities of the excited state P_e are obtained:

$P_e = 0$ (state $|1\rangle$) if the delay between the two Ramsey pulses is zero.

$P_e = 0.5$ (superposition of the same weight of states $|1\rangle$ and $|0\rangle$) if the delay between the two Ramsey pulses corresponds to a quarter period of the Ramsey fringes (which corresponds on the Bloch sphere to the application of the second Ramsey pulse when crossing the direction y of the Bloch sphere).

$P_e = 1$ (state) if the delay between the two Ramsey pulses corresponds to half the period of the Ramsey fringes or to applying the second Ramsey pulse when crossing the -x direction of the Bloch sphere.

4

Reliability-based Design Optimization of Structures

The objective of optimizing mechanical structures is to select the best possible design in terms of cost and quality. One of the major stakes of reliability-based design optimization is to monitor, predict and detect the failure modes of the systems under study. This chapter presents design optimization and reliability methodologies that include mechanical uncertainties. These approaches contribute to the competitiveness of companies in the automotive, aeronautics, civil engineering and defense sectors. Various applications highlight the reliability-based design optimization method.

4.1. Introduction

Traditionally, the design engineer optimizes the design of the mechanical structures of a system from the results of successive tests. On the basis of experience, field returns and know-how, he/she designs a first version, and then verifies by calculation whether it meets the specification requirements. If this is not the case, he/she modifies the design until the functional requirements and constraints are respected. In most cases, several iterations are necessary. This method is expensive in terms of R&D leading time and prototype building cost. However, in this approach, the designer does not care about the degree of precision of the mechanical characteristics of the materials, the geometry and the loading. To improve the efficiency of the design process, digital modeling and optimization software are used to study the various possible solutions. Design optimization focuses on obtaining performance and minimizing cost. However, the data of the degradation of the reliability caused by the use cycles and the operational conditions should be taken into account.

This chapter presents a reliability-based design optimization method. This method guarantees a trade-off between the cost of the system design and the customer's expected performance under the intended conditions of use. It is based on the uncertainties of the design elements, the constraints and the simultaneous resolution of two problems: optimizing the cost and performing the required functions for a sufficiently high operation probability under the use conditions.

4.2. Deterministic optimization

The mathematical formulation of a deterministic optimization is written as [ELH 13]:

$$\begin{aligned} \text{min} \quad &: f(x) \\ \text{under} &: g_1(x) = \left(G_1(x) - G_1^t\right) \geq 0 \\ &: g_2(x) = \left(G_2(x) - G_2^t\right) \geq 0 \end{aligned} \qquad [4.1]$$

where $\{x\}$ represents the deterministic vector of the design parameters.

The deterministic optimization method is based on analyzing the geometric and material properties and the loads. This method provides a detailed behavior of the structure under study. Figure 4.1 shows a deterministic optimization based on the safety factor. But this approach has its limits. Variability in structural properties as well as in modeling approximations may decrease the probability that the mechanical system operates properly under operational conditions. It is therefore necessary to analyze the effects of uncertainties in the parameters and approximations on the reliability of the product under study.

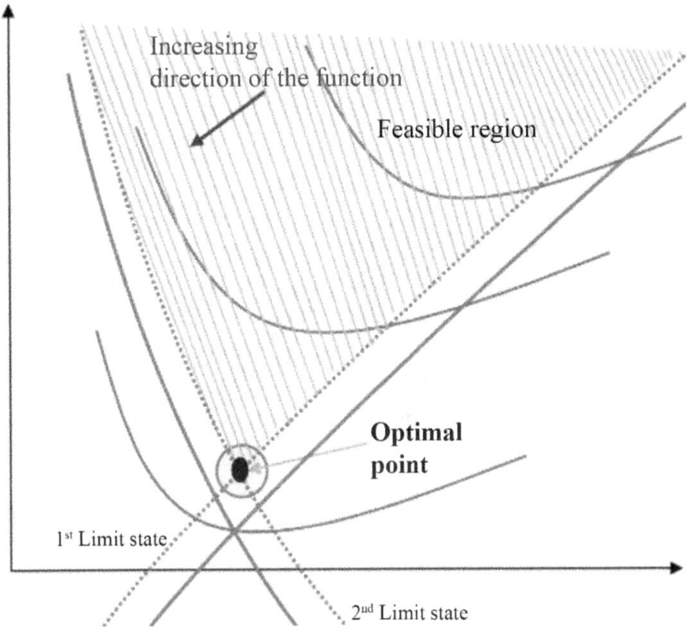

Figure 4.1. *Deterministic optimization based on reaching a safety factor level. For a color version of this figure, see www.iste.co.uk/dahoo/metrology2.zip*

4.3. Reliability analysis

The principles of reliability that are applied to structural mechanics problems are reviewed in [ELH 13].

A random vector of design variables denoted by $\{Y\}$ is considered at first; the realizations of $\{Y\}$ are denoted as $\{y.\}$ Reliability is expressed by the probability of success of a scenario represented by a limit state $G(x, y)$ that is a function of the random variables $\{y\}$ and the deterministic variables x:

$$Reliability = 1 - P_f = 1 - Prob\left[G(\{x\},\{y\}) \leq 0\right] \qquad [4.2]$$

$G(x, y) = 0$ defines the limit state, $G(x, y) > 0$ scans the safety state domain and $G(x, y) < 0$ likewise scans the failure state domain. A reliability index β is introduced as a measurement of the reliability level. The computation of an exact and invariant index implies that it is defined in a new space of statistically independent Gaussian variables $\{u\}$, with a mean and standard deviation equal to unity (Figure 4.2), which is different from the space of physical variables $\{y\}$. An iso-probabilistic transformation is defined for the transformation from one space to the other:

$$u = T(x; y)$$

In this standardized space, the reliability index β represents the minimum distance between the origin of the space and the limit state function $H(x, u) = G(x, y)$. The point that is closest to the origin is called the design point. The reliability index is calculated by an optimization procedure under the constraint that it is at the border of the failure domain:

$$\begin{aligned} &\beta = \min\left(\sqrt{\{u\}^T\{u\}}\right) \\ &\textit{under } H\left(\{x\},\{u\}\right) \leq 0 \end{aligned} \qquad [4.3]$$

where $\{u\}$ is the modulus of the vector in the normal space, as measured from the origin.

The solution of the problem described by equations [4.3] is called the determination of the design point P^*. This solution is subject to the difficulties of nonlinear programming such as existence of local minima, gradient approximation and computation time. Although equations [4.3] can be solved by any suitable optimization method, specific algorithms have been developed to solve this reliability problem. In the book [ELH 13], four criteria are used to compare these algorithms: generality, robustness, efficiency and capacity. Five algorithms are thus recommended to assess the reliability of structures: sequential quadratic programming (SQP), the modified Rackwitz–Fiessler (RF) method, the projected gradient method, the augmented Lagrangian method and the penalty method. However, in nonlinear finite element analysis, the projected gradient method is less efficient.

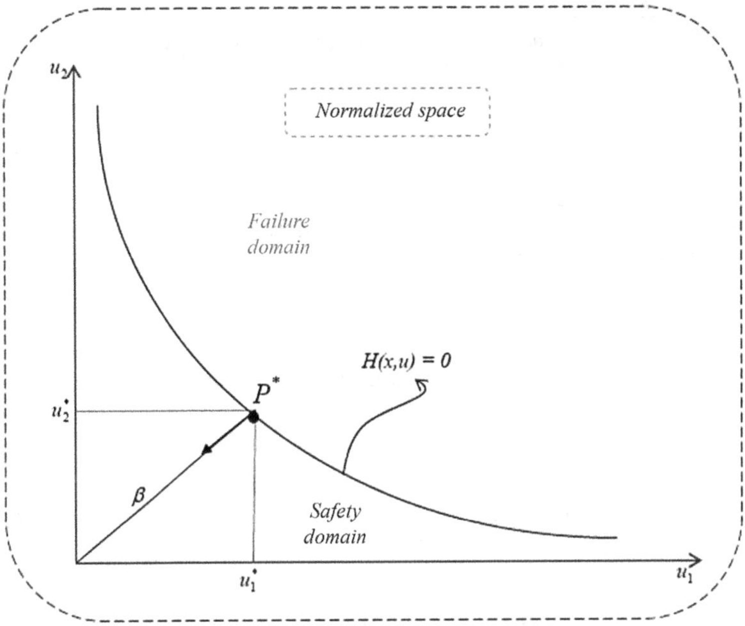

Figure 4.2. *Normal physical space. For a color version of this figure, see www.iste.co.uk/dahoo/metrology2.zip*

4.3.1. *Optimal conditions*

The optimization problem [4.3] is equivalent to the problem posed by the constrained minimization of a Lagrangian:

$$L(u,\lambda_H) = u^T.u + \lambda_H\, H(x,y) \qquad [4.4]$$

with λ_H representing the Lagrangian multipliers.

The optimal conditions for the Lagrangians are:

$$\frac{\partial L}{\partial u_j} = 2u_j + \lambda_H \frac{\partial H}{\partial u_j} = 0 \qquad [4.5]$$

$$\frac{\partial L}{\partial \lambda_H} = H(x,u) = 0 \qquad [4.6]$$

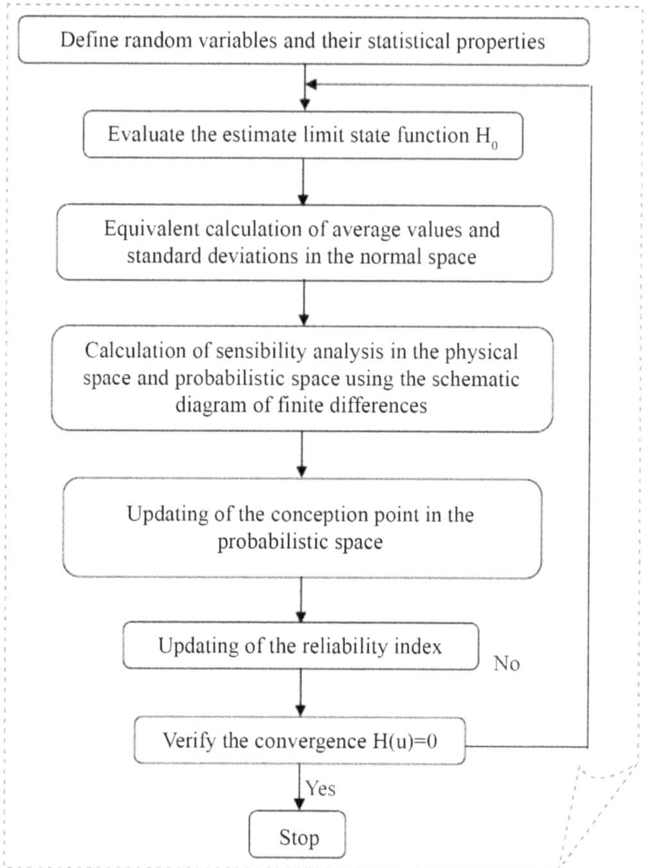

Figure 4.3. *Reliability index assessment process*

This method (Figure 4.3) consists of evaluating the Lagrangian derivatives in the normalized space. More particularly, the limit state function $H(x, u)$ is unknown until finite element analysis evaluation of the latter is obtained. However, such an analysis requires a long computation time. This is particularly the case for nonlinear and transient problems:

Moreover, the normalized gradient analysis $\frac{\partial H}{\partial u_j}$ cannot be calculated directly because mechanical analysis is performed in the physical space, and not in the standard normalized space. The calculation of the normalized gradient is carried out by applying the chain rule on the gradient $\frac{\partial G}{\partial y_k}$ in the physical space:

$$\frac{\partial H}{\partial u_j} = \frac{\partial G}{\partial y_k} \cdot \frac{\partial T^{-1}(x,u)}{\partial u_j} \quad [4.7]$$

These derivatives are generally obtained by finite difference methods, which require a long computation time. Integrating reliability analysis into design optimization engineering is called reliability-based design optimization (RBDO).

4.4. Reliability-based design optimization

The objective of designing under uncertain parameters is to find the optimum trade-off between the reliability level of the structure and its optimal design cost. Figure 4.4 shows the comparison between the optimal solutions of the deterministic design optimization (DDO) method and the reliability-based design optimization (RBDO) method. The solution of the DDO method is found in the vicinity of point A that is at the border of a limit state which may lead to an erroneous solution. The solution of the RBDO method is found in the feasible space domain around point B.

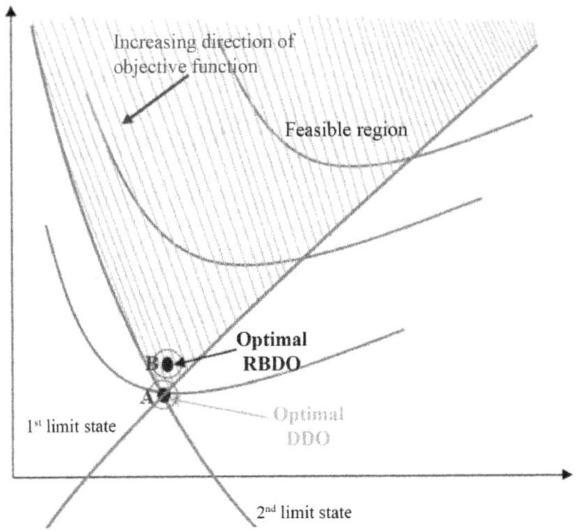

Figure 4.4. *Comparison between the solutions of the RBDO and DDO methods. For a color version of this figure, see www.iste.co.uk/dahoo/metrology2.zip*

4.4.1. *The objective function*

Several objective functions can be used in the RBDO method. These functions include cost and utility functions that should be minimized or maximized. The optimal life cost and the optimal utility can be calculated [KHA 04].

4.4.2. *Taking into account the total cost*

If reliability aspects are not taken into account in design optimization, the solutions will not be economic, as an increase in the failure rate in operation may introduce a higher cost than the expected savings. Indeed, the mechanical structure design codes do not necessarily provide homogeneous reliability levels and solutions that respect regulations.

The expected total cost (C_T) of a structure is obtained by the linear combination of the initial cost, the cost of failure and the maintenance cost (Figure 4.5), which can be written as follows:

$$C_T = C_c + C_f P_f + \sum_r C_{I_r} P_{I_r} + \sum_s C_{M_s} P_{M_s} \qquad [4.8]$$

where C_c is the cost of constructing the structure, C_f is the cost of failure due to direct and indirect damage to a structural component, C_{Ir} is the cost of inspection, C_{Ms} is the cost of maintenance and repair, P_f is the probability of failure, P_{Ir} is the probability of non-failure until detection by the rth inspection and P_{Ms} is the probability of repair.

The objective of the reliability-based optimization method is to minimize the total cost C_T of the structure. Equation [4.8] is not relevant because of the difficulty of assessing the cost of failure C_f (especially when it comes to immaterial damage). Hence, the optimization problem becomes more significant when the initial cost, represented by the objective function $f(x)$ under the constraint of satisfying a target reliability level $\beta \geq \beta_t$, is damped. The focus is on minimizing the initial cost and the cost of failure.

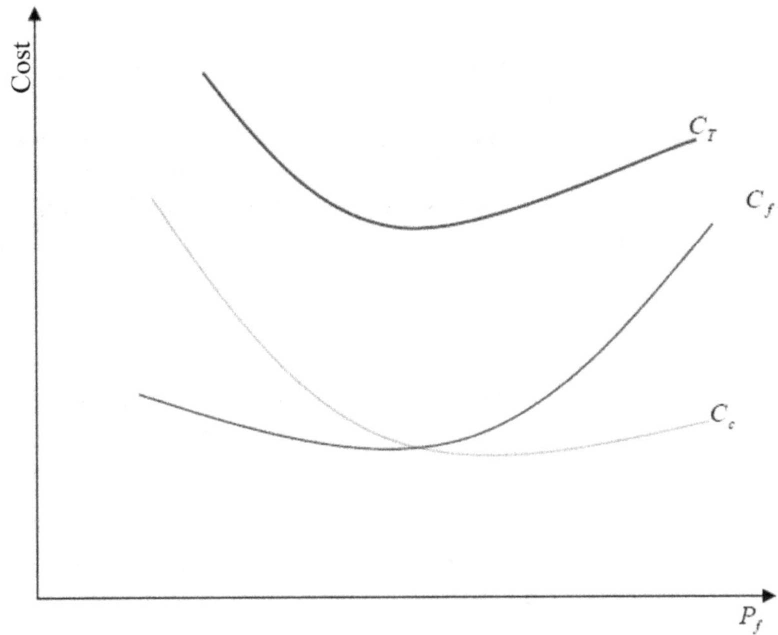

Figure 4.5. *Total cost (C_T), cost of failure (C_f) and initial cost of the structure (C_c) as a function of the probability of failure (P_f). For a color version of this figure, see www.iste.co.uk/dahoo/metrology2.zip*

4.4.3. *Design variables*

The design variables [KHA 08] can be classified into scale variables, configuration-based variables, topological variables and material variables. These variables can be continuous or discrete.

4.4.4. *Response of a system by RBDO*

Generally, the RBDO method is applied to the structures that have a linear static behavior, although some studies have been devoted to dynamics [MOH 10].

4.4.5. *Limit states*

On the one hand, most techniques use sequential limit states (Figure 4.4) and do not take into account the interactions between limit states. On the other hand, the RBDO method is a systemic approach and takes into account all the appropriate limit states of a structural system and their interactions. The RBDO method leads to a structure that will meet the reliability requirements for a given limit state.

4.4.6. *Solving methods*

In the last decade, several numerical RBDO optimization techniques have been developed. The proposed algorithms are presented in the book [ELH 13].

To solve an RBDO problem, optimization algorithms can be coupled with a specific reliability software [KAY 94, ELH 13]. However, most of the solving methods used in RBDO have been applied to industrial structural systems of small or moderate size.

Among the recently proposed methods are the RIA (Reliability Index Approach) and the PMA (Performance Measure Analysis) methods. As with all the classical methods of reliability optimization, these methods are based on solving in two spaces: the normalized space of the random variables and the physical space of the design variables. Another method that couples these two spaces into a single hybrid space has been developed. It has been demonstrated that this method is as efficient as the classical RBDO approaches [KHA 14]. Other methods based on the main advantages of the hybrid space have been developed [MOH 10].

4.5. Applications

4.5.1. *Application on a bending beam*

A cantilever beam that is free at one end is considered (Figure 4.6). Its parameters are the following: length 1 m, Young's modulus E_0=2.1.10^{11} N.m^{-2} and density $\rho = 7{,}800$ kg/m^3.

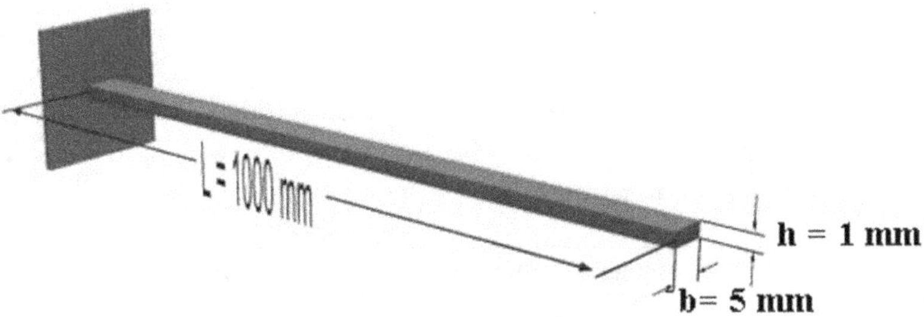

Figure 4.6. *Cantilever beam in free bending mode. For a color version of this figure, see www.iste.co.uk/dahoo/metrology2.zip*

The objective is to minimize the volume of the beam that is subjected to the stress of the natural bending frequency. The system must respect a predetermined reliability target. The target index is based on statistical reliability studies; however, for several technical applications, the target reliability index is set as $\beta_t = 3.8$. The displacement is usually given as a function of frequency and is often reduced to the first resonant frequency. In order to guarantee the required level of safety, the performance of the structure under study must be in a feasible domain outside the interval $[f_a, f_b]$ that is centered on the resonant frequency. In this example, RBDO is compared with a DDO calculation. The complete model has 16 degrees of freedom.

4.5.1.1. *RBDO method*

The RBDO hybrid problem is defined by:

$$\text{minimizing: } F(\{x\},\{y\}) = f(\{x\}) \,.\, d_{\beta 1}(\{x\},\{y\}) \,.\, d_{\beta 2}(\{x\},\{y\})$$

$$\text{subject to: } f(L,b,h) - f_c = 0\,,\ \beta_1(\{x\},\{y\}) \le \beta_t \text{ and } \beta_2(\{x\},\{y\}) \le \beta_t\ .$$

with $f_c = 10$ Hz

where the variables L, b and h are grouped in the random vector $\{Y\}$ having as means the nominal values m_L, m_b and m_h, which are grouped in a vector $\{X\}$, and as standard deviations *0.1; 0.0005; 0.0001.*

Table 4.1 summarizes the results of the RBDO and DDO methods for a complete structure model and a condensed structure model. The use of the sub-structuring technique has drastically reduced the number of degrees of freedom and the calculation time. The frequency band that verifies the level of reliability requested is the same for the two methods. The RBDO model is an efficient tool for controlling this frequency band.

Variables	Complete structure model			Condensed structure model		
	Design point (A)	Optimal solution	Design point (B)	Design point (A)	Optimal solution	Design point (B)
L (mm)	792.64	913.82	792.64	790.79	955.49	790.79
h (mm)	3.8299	3	5.1	3.5	4.2242	5.1009
B (mm)	0. 71588	1.0355	0.9	0.70857	1.0189	0.89997
β	3.8	–	3.8	3.8	–	3.8
Frequency (Hz)	9.1889	10	11.5522	9.2	10.02	11.5532
Volume (mm^3)	–	2.8388e-3	–	–	4.1125e-3	–

Table 4.1. *RBDO results for a calculation before and after condensing the model*

4.5.2. *Application on a circular plate with different thicknesses*

The circular plate under study (Figure 4.7) is made of a copper alloy where the thickness varies from the center to the perimeter at various radii. The geometrical and mechanical characteristics of the plate are given by:

Young's modulus: $E = 1.06\ 10^{11}\ N.m^{-2}$;

Density: $\rho = 8{,}400\ kg.m^{-3}$;

Thickness: $h_1 = 3$ mm, $h_2 = 0.5$ mm, $h_3 = 1.5$ mm;

Radius: $R_1 = 23$ mm, $R_2 = 27.5$ mm, $R_3 = 38$ mm, $R_4 = 50$ mm.

The objective is to minimize the volume that is subjected to the stress of the natural resonant frequency so that the first resonant frequency of the structure is higher than 500 Hz. The system must respect a predetermined reliability target. The target index is usually based on statistical reliability studies; however, for this technical application, the target reliability index is set as $\beta_t = 3.8$.

The graph of the amplitude of the displacement versus the frequency is shown in Figure 4.9. In order to guarantee the required level of safety, the performance of the structure under study must be in a feasible domain outside the interval $[f_a, f_b]$ that is centered on the natural resonant frequency.

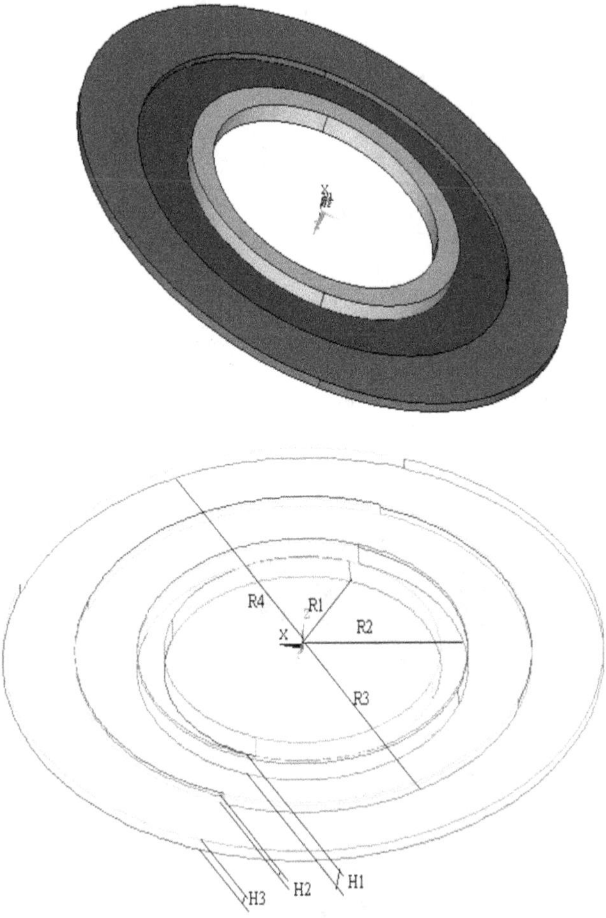

Figure 4.7. *3D circular plate. For a color version of this figure, see www.iste.co.uk/dahoo/metrology2.zip*

The DDO and RBDO methods are applied to a complete and condensed model.

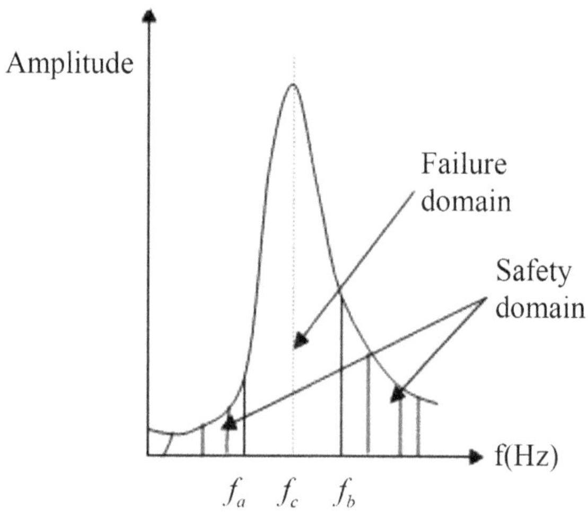

Figure 4.8. *Amplitude of the displacement versus frequency; the area corresponding to the safety level is in a feasible domain outside the interval [f_a,f_b] that is centered on the resonant frequency f_c*

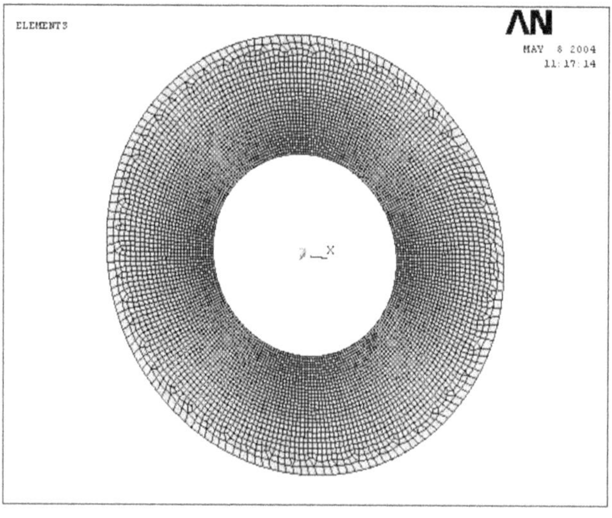

Figure 4.9. *Model of the circular plate. For a color version of this figure, see www.iste.co.uk/dahoo/metrology2.zip*

The complete model of the structure (Figure 4.9) has 40,224 active degrees of freedom, and the condensed model has 1,000 degrees of freedom.

4.5.2.1. *Optimization problem:*

The volume of the stressed circular plate is minimized, as it is subject to the natural resonant frequency:

$$\text{min: } volume(R1, R2, R3, H1, H2, H3)$$

$$\text{under: } f(R1, R2, R3, H1, H2, H3) - f_c\,S_f = 0 \text{ and } f_c\text{=}500\ Hz,\ S_f\text{=}1.2$$

4.5.2.2. *Reliability analysis of the optimal design:*

For a normal distribution, the normalized variable u is written as:

$$u_i = \frac{x_i - m_{xi}}{\sigma_{xi}}$$

with $\{x_i\}$= $\{R1, R2, R3, R4, H1, H2, H3\}$.

To calculate the reliability index introduced by Hasofer–Lind [25], two subproblems are formulated:

$$\beta = \min\ d_1(\{u\}) = \sqrt{\sum_1^m u_j^2} \quad \text{under: } f(R1, R2, R3, H1, H2, H3) - f_c\,S_f = 0 \quad \text{with}$$

f_c = 500 Hz.

4.5.2.3. *RBDO*

The classical RBDO method leads to a low stability of convergence, but the hybrid method makes it possible to couple reliability analysis and optimization problems [KHA 02].

The hybrid RBDO problem is written as:

$$\text{min: } F(\{x\}\{y\}) = f(\{x\}) \,.\, d_\beta(\{x\},\{y\})$$

$$\text{Under: } f(R1, R2, R3, R4, H1, H2, H3) - f_c = 0, \; \beta_1(\{x\},\{y\}) \leq \beta$$

where the variables *R*1, *R*2, *R*3 *R*4, *H*1, *H*2 and *H*3 are grouped in the random vector {Y} having as means the nominal values m_{R1}, m_{R2}, m_{R3}, m_{R4}, m_{H1}, m_{H2} and m_{H3}, which are grouped in a vector {X}, with standard deviations equal to $0.1.m_{X_i}$.

	DDO		RBDO	
Variable	**Design point**	**Optimal point**	**Design point**	**Optimal point**
R1 (mm)	23.687	22.96	22.845	24.892
R2 (mm)	26.858	29.23	27.445	27.072
R3 (mm)	37.584	39.98	37.962	36.906
R4 (mm)	51.376	47.01	49.785	48.322
H1 (mm)	2.558	2.51	3.194	2.599
H2 (mm)	0.587	0.41	0.496	0.632
H3 (mm)	1.51	1.42	1.413	1.74
β	1.7	–	3.8	–
Frequency (Hz)	500	600.26	500	693
Volume (mm^3)	–	6266.9	–	7490.2089

Table 4.2. *DDO and RBDO results*

The results obtained by the DDO and RBDO hybrid methods are presented in Tables 4.2 and 4.3. The DDO method does not provide an acceptable estimate of the reliability of the structure. However, the hybrid RBDO method overcomes this problem. Coupling the hybrid method with the model reduction method, which has only 1,000 active degrees of freedom, provides the same results with a 1% error (negligible) for a reduced computation time.

	RBDO (complete structure)		RBDO (condensed structure)	
Variable	Conception point	Optimal point	Conception point	Optimal point
R1 (mm)	22.845	24.892	22.519	24.986
R2 (mm)	27.445	27.072	27.407	26.966
R3 (mm)	37.962	36.906	37.794	37.083
R4 (mm)	49.785	48.322	49.994	48.135
H1 (mm)	3.194	2.599	2.646	3.277
H2 (mm)	0.496	0.632	0.50	0.633
H3 (mm)	1.413	1.74	1.446	1.74
β	3.8	–	3.8	–
Frequency (Hz)	500	693	500	702
Volume (mm^3)	–	7490.2	–	7487.8

Table 4.3. *RBDO results for a calculation before and after model condensation*

4.5.3. *Application: hook A*

A comparative study of the DDO and RBDO methods is carried out through this application. The design of hook A is shown in Figure 4.10. The various predefined geometric dimensions and material properties are presented in Table 4.4.

Setting		Value
Young's modulus (MPa)		2.1×106
Volume density		78×10-7
Poisson's coefficient		0.3
D1	(mm)	20
D2	(mm)	15
D3	(rad)	2.5
D4	(rad)	35
D5	(mm)	5
D6	(mm)	1
Rc	(rad)	0.5
R0	(mm)	3

Table 4.4. *Initial parameters*

The hook loading environment consists of the welded stress around the upper circumferences, as well as of the up and down pressure applied to the lower half of the lower hole $P = 150$ MPa.

The objective is to optimize the hook design under an allowable stress $\sigma_{ad} = 235\,MPa$ and a target reliability index $\beta_c = 3.8$, and then optimize the design when subject to another frequency constraint.

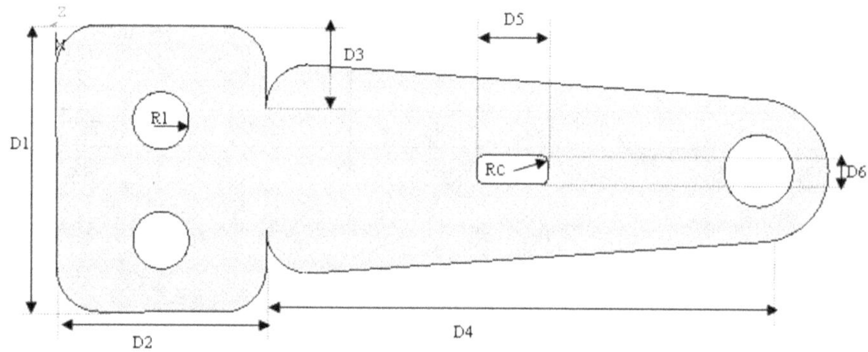

Figure 4.10. *Dimensions of the section of the hook under study. For a color version of this figure, see www.iste.co.uk/dahoo/metrology2.zip*

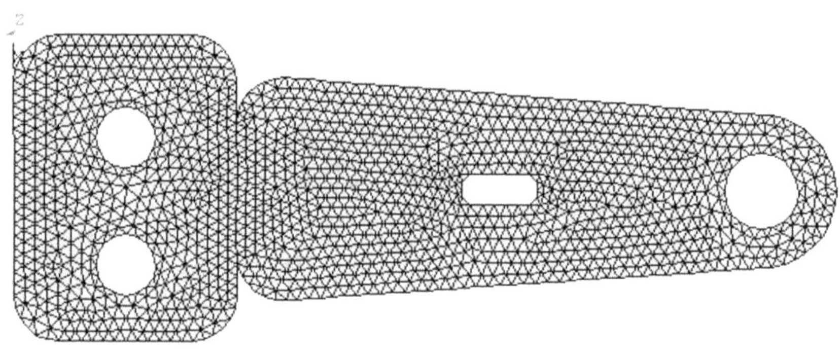

Figure 4.11. *Finite element model of the hook. For a color version of this figure, see www.iste.co.uk/dahoo/metrology2.zip*

4.5.3.1. *Problem 1: the case of a single constraint*

4.5.3.1.1. First study by the DDO method

To perform the DDO method, various global safety factors $S_f = 1.5, 1.2, 1.1$ are proposed for the applied load and a sensitivity analysis of the parameters is performed in parallel.

Formulation of the DDO problem: min: $Volume(D1, D2, D3, D5, D6, Rc, R0)$ (II-21a)

under: $\sigma_{eqv}(D1, D2, D3, D5, D6, Rc, R0) - \sigma_{ad} = 0$

and $\sigma_{ad} = 235\, MPa$.

Reliability analysis of the optimal solution: for a normal distribution, the normalized variable u has the following form:

$$u_i = \frac{x_i - m_{x_i}}{\sigma_{x_i}} \qquad [4.9]$$

where $\{x_i\} = \{D1, D2, D3, D5, D6, Rc, R0\}$.

To calculate the reliability index presented by Hasofer–Lind [HAS 74], the problem is defined by:

$$\beta = \min d_1(\{u\}) = \sqrt{\sum_1^m u_j^2} \qquad [4.10]$$

$$under: \sigma_{eqv}(D1, D2, D3, D5, D6, Rc, R0) - \sigma_{ad} = 0$$

with $\sigma_{ad} = 235\, MPa$.

Parameters	Design point	Optimal point
D1 (mm)	15.215	14.552
D2 (mm)	10.125	9.3327
D3 (mm)	2.3730	1.8710
D5 (mm)	21.004	17.153
D6 (mm)	1.8632	2.4532
Rc (rad)	1.0688	1.4214
R0 (mm)	1.1603	1.6066
σ_{eqv} (MPa)	235.05	156.73
Volume (mm^3)	–	563.40
Reliability index	5.7289	
Safety factor	1.5	

Table 4.5. *DDO results for a safety factor of 1.5*

Parameter	Design point	Optimal point
D1 (mm)	15.119	14.861
D2 (mm)	10.021	9.3502
D3 (mm)	2.3664	1.8965
D5 (mm	22.029	22.155
D6 (mm)	1.8476	2.5404
Rc (rad)	1.0303	1.3042
R0 (mm)	1.1744	1.5083
σ_{eqv} (MPa)	235.10	195.89
Volume (mm^3)	–	502.63
Reliability index	4.8413	
Safety factor	1.2	

Table 4.6. *DDO results for a safety factor of 1.2*

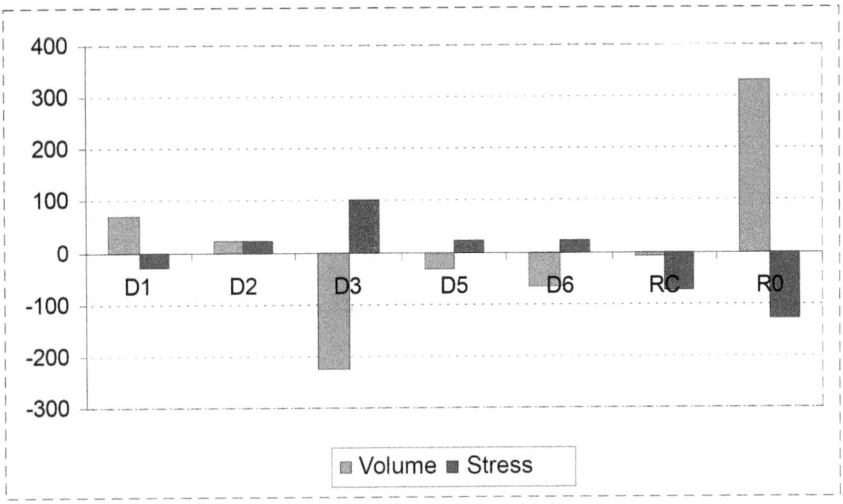

Figure 4.12. *Sensitivity analysis of the parameters D1, D2, D3, D5, D6, Rc, R0 with respect to the volume and stress* $\frac{\partial}{\partial x}$ *and* $\frac{\partial G}{\partial x}$. *For a color version of this figure, see www.iste.co.uk/dahoo/metrology2.zip*

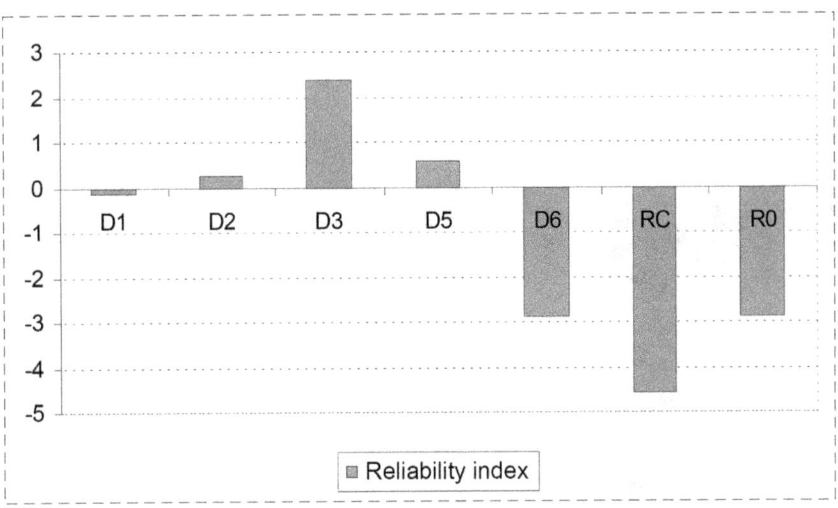

Figure 4.13. *Sensitivity analysis of the parameters D1, D2, D3, D5, D6, Rc, R0 with respect to the reliability index* $\frac{\partial \beta}{\partial x_i}$. *For a color version of this figure, see www.iste.co.uk/dahoo/metrology2.zip*

Parameter	Design point	Optimal solution
D1 (mm)	15.160	14.910
D2 (mm)	10.008	10.208
D3 (mm)	2.3775	2.4092
D5 (mm)	21.994	22.996
D6 (mm)	1.8546	1.8936
Rc (rad)	1.0118	0.84577
R0 (mm)	1.1752	1.3853
σ_{eqv} (MPa)	235	213.75
Volume (mm^3)	–	434.75
Reliability index	2.54	
Safety factor	1.1	

Table 4.7. *DDO results for a safety factor of 1.1*

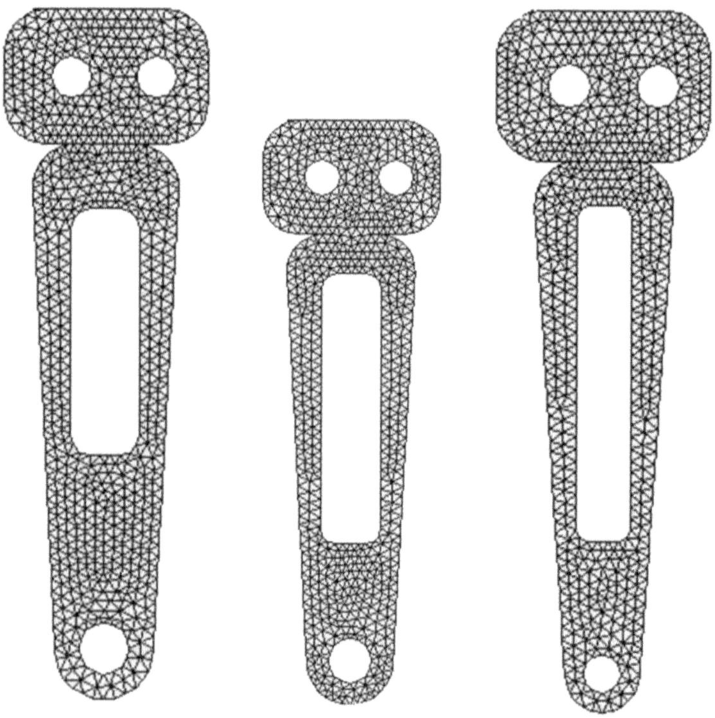

Figure 4.14. *Various shapes of the optimal volume solution for the safety factors 1.5, 1.2 and 1.1. For a color version of this figure, see www.iste.co.uk/dahoo/metrology2.zip*

Safety factor	1.5	1.2	1.1
Optimal volume	563.40	502.63	434.75
Volume reduction	73.5%	76.4%	79.6%

Table 4.8. *Comparison of the results obtained by the DDO method for the various safety factors*

Discussion

Tables 4.7 and 4.8 summarize the results of the DDO method for the various assumed safety factors. Figures 4.14 and 4.15 show the influence of the various design parameters, and Figure 4.16 shows the optimal shapes of the structure with respect to the different assumed safety factors.

By comparing these results, a volume reduction of 73.5% is obtained for a safety factor of $S_F = 1.5$ and 76.4% for a safety factor of $S_F = 1.2$. For these results, an assessment of the reliability of the solution is carried out: the reliability indices $\beta_{1.5} = 5.7$ and $\beta_{1.2} = 4.8$ are very high compared to the target reliability index $\beta_t = 3.8$. From these results, a cost reduction is obtained by reducing the safety factor to $S_F = 1.1$. In this case, the volume reduction is 79.6% with a reliability level of $\beta_{1.1} = 2.5 \leq \beta_t$.

4.5.3.1.2. Second study by the RBDO method

The DDO approach does not respect the reliability level, and the classical RBDO method leads to poor convergence stability because of the nested solution of the two optimization problems. However, the hybrid method makes it possible to couple the reliability analysis and optimization solving methods.

Formulation of the problem

The formulation of the hybrid RBDO problem is written as:

$$\min : F(\{x\},\{y\}) = f(\{x\}) \times d(\{x\},\{y\})$$
$$under : \sigma_{eqv}(D1, D2, D3, D5, D6, Rc, R0) - \sigma_{ad} = 0 \qquad [4.11]$$
$$and\ \beta(\{x\},\{y\}) \geq \beta_t$$

where $D1$, $D2$, $D3$, $D5$, $D6$, Rc and $R0$ are grouped in the random vector {Y} having as means the nominal values m_{D1}, m_{D2}, m_{D3}, m_{D5}, m_{D6}, m_{Rc} and m_{R0}, which are grouped in a vector {X}, with a standard deviation of 0.1.

Parameter	Design point	Optimal solution
D1 (mm)	15.994	15.129
D2 (mm)	10.515	11.857
D3 (mm)	2.4546	2.0451
D5 (mm	21.543	21.167
D6 (mm)	1.9851	1.8881
Rc (rad)	0.84454	0.90157
R0 (mm)	1.2200	1.6739
Stress (MPa)	235.07	151.75
Volume (mm^3)	–	666.02
Reliability index	3.8	

Table 4.9. *RBDO results*

Table 4.9 summarizes the results of the RBDO method: The volume reduction is 68.7% for a reliability level equal to 3.8. The shape of the optimal solution and its von Mises stress are shown in Figure 2.16.

The RBDO approach respects the safety requirements, which proves that this approach leads to an efficient coupling between the reliability analysis and the optimization method.

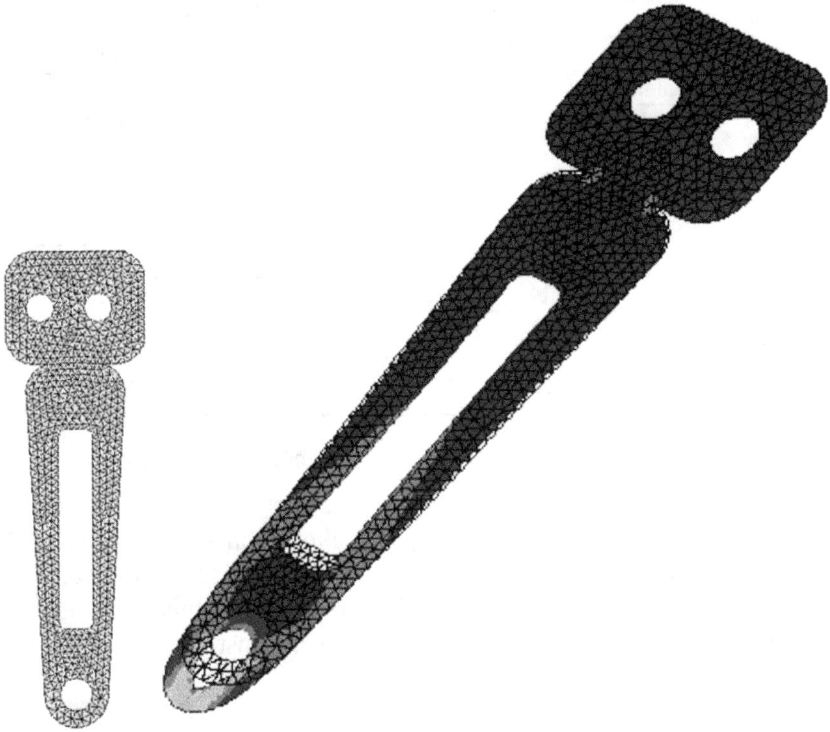

Figure 4.15. *Optimal solution shape of volume and von Mises stress. For a color version of this figure, see www.iste.co.uk/dahoo/metrology2.zip*

4.5.3.2. *Problem 2: the case of multiple mechanical stresses*

In this part of the study, a new constraint is added. The formulation of the problem is written as:

$$\begin{aligned} &\min : F(\{x\},\{y\}) = f(\{x\}) \times d(\{x\},\{y\}) \\ &under : \sigma_{eqv}(D1, D2, D3, D5, D6, Rc, R0) - \sigma_{ad} = 0 \\ &\qquad f(D1, D2, D3, D5, D6, Rc, R0) \geq f_{ad} \\ &and\ \beta(\{x\},\{y\}) \geq \beta_t \end{aligned} \qquad [4.12]$$

with $\sigma_{ad} = 235\,MPa$ and $f_{ad} = 270\,Hz$.

Parameter	Design point	Optimal solution
D1 (mm)	15.351	15.082
D2 (mm)	10.412	11.641
D3 (mm)	2.3564	1.9267
D5 (mm)	21.454	22.096
D6 (mm)	1.9409	1.8741
Rc (rad)	0.90430	0.99322
R0 (mm)	1.1212	1.5723
Stress (MPa)	235.09	206.087837
Frequency (Hz)	270.07	281.95
Volume (mm^3)	–	634.334496
Reliability index	3.8002	

Table 4.10. *Results of the RBDO method for several limit states*

The volume reduction is 70.21%, and the mechanical and reliability constraints are respected (see Table 4.10). Figure 4.16 shows the von Mises stress *t* and the first mode along the X axis of the optimal solution. The RBDO method respects the safety and mechanical constraints.

Figure 4.16. *The von Mises stress and the first mode along the X axis of the optimal solution. For a color version of this figure, see www.iste.co.uk/dahoo/metrology2.zip*

4.5.4. *Application: optimization of the materials of an electronic board*

Embedded electronic systems must respect their performance specifications for increasingly longer service periods. Therefore, it is important to verify the performance of these systems and their reliability. The increase in system complexity due to the integration of various technologies has a strong effect on the functional safety and implicitly on the reliability of these systems. Many authors have demonstrated this correlation [ELH 13]. Indeed, the failure rate increases with the miniaturization of the electronic systems, with the increasing density of the components on the electronic boards, as well as with the heterogeneous mechanical and thermal properties of the composite materials. The necessity of reducing the volume and weight of the electronic equipment has led printed circuit board (PCB) manufacturers to implement new technologies.

The primary market for PCB manufacturers is a 1.5 mm-thick double- or single-sided copper composite laminate. A smaller share of the market concerns multilayer circuits composed of thinner copper layers and flexible deformable circuits. Most PCB circuits are made up of brominated or non-brominated solid resins (corresponding to the classes G10 and FR4 of the standard defined by the National Electrical Manufacturer Association (NEMA)), which are mixed with an accelerated hardener (dicyandiamide) in the liquid phase. The circuits of the higher electrical classes (G11 and FR5) made up of liquid bisphenol A epoxy resins with aromatic amines are less frequent.

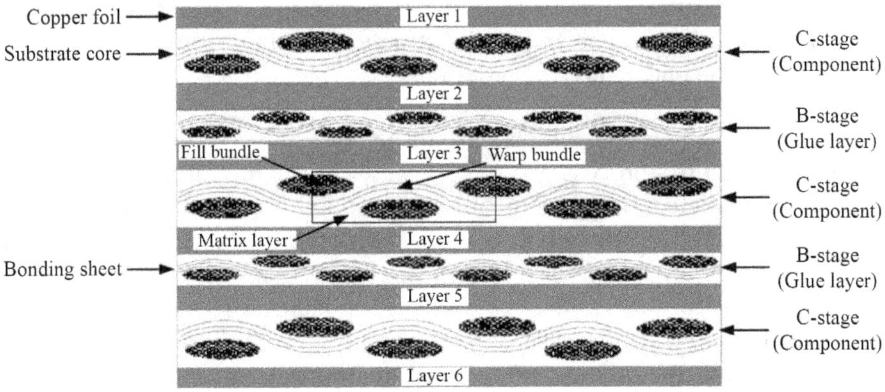

Figure 4.17. *Schematic cross-section of a six-layer printed circuit r*

Although the PCB is a structured composite, system designers consider it as a component such as integrated circuits, resistors, capacitors or others. A PCB consists of a substrate that is generally composed of composite laminated materials (FR2, FR3, FR4, etc.), which are a mixture of a fibrous reinforcement (glass fiber, aramid fiber, etc.), an organic or inorganic matrix (epoxy resin, glass resin, etc.) and copper layers or tracks (Figures 4.17 and 4.18).

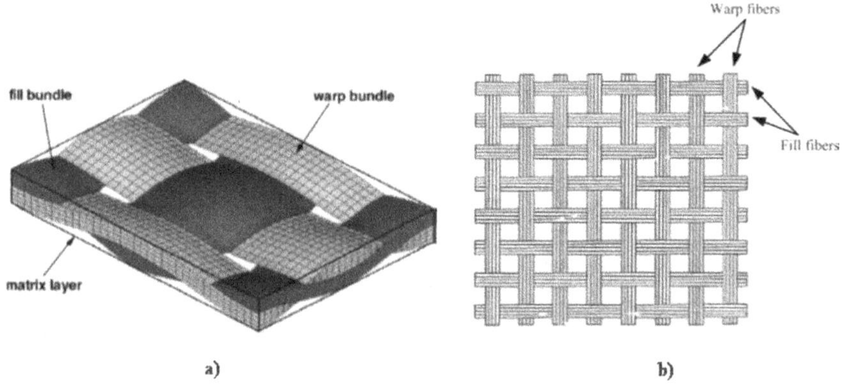

Figure 4.18. *Architecture of the fiber-reinforced PCB: a) overview, b) detail of the fiber fabric*

The choice of these composite materials is due to their low cost, good temperature resistance, excellent adhesion to copper and electromechanical behavior. To improve the mechanical performance of the PCB structure, it is necessary to adapt the architecture of the fiber preform, i.e. to adjust the volume rate of the fibers and the orientation angles of the plies. Once the structural optimization process has been completed, the designer can be provided with the parameters of the structure to study the definition and feasibility of the manufacturing process.

FR4 materials are manufactured using either wet or dry processing techniques. In a wet process, the mixture between the reinforcements and the matrix is completed on the semi-finished product (prepreg) before the final shaping phase. This procedure provides a better impact resistance than injecting a resin system into a dry reinforcement thanks to the presence of thermoplastics in the resin. The prepregs obtained by drying are then subjected to a 20 MPa pressure, at a temperature ranging from 160 to 200°C. In a dry process, the mixture is produced by injection or infusion during the shaping phase of the finished part. This method reduces storage costs and increases the thicknesses of the reinforcements.

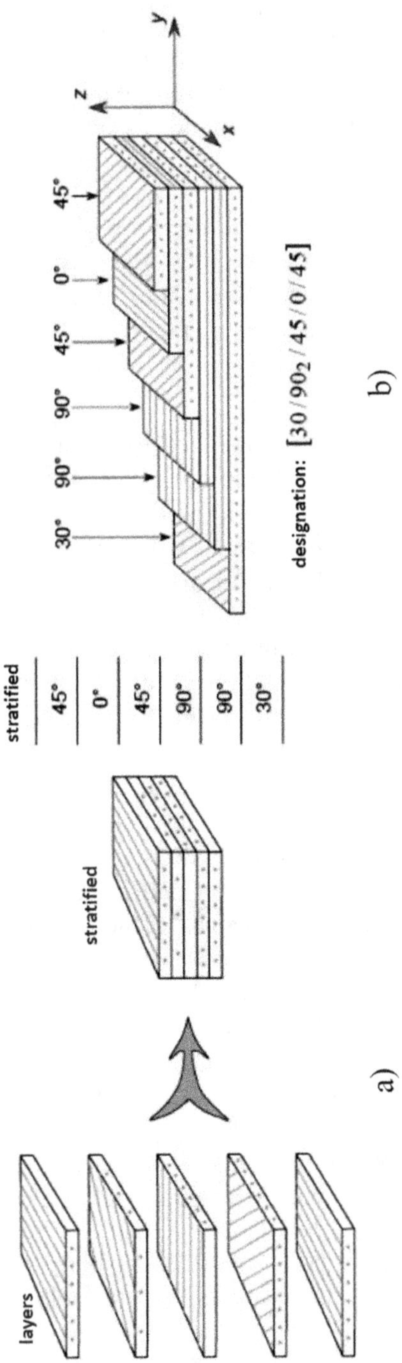

Figure 4.19. *a) A laminate structure; b) types of laminates. For a color version of this figure, see www.iste.co.uk/dahoo/metrology2.zip*

One of the main challenges of the composite materials industry is the development of new processes that combine the performance of prepregs with the economic advantages of dry processes. In this case, the LCM (Liquid Composite Molding) types of processes seem to be a promising option.

Laminates based on unidirectional fibers or matrices constitute a basic type to which, in theory, any other type of laminates can be reduced. These laminates are composed of layers of unidirectional fibers or matrices, the direction of which is offset in each layer (Figure 4.19).

4.5.4.1. *Optimization issues*

Reducing the dimensions of embedded electronic systems requires an optimal design of the PCB, leading to a mass reduction and an increase in the resonant frequencies of the natural modes of vibration. As the PCB is made up of various types of materials, a structural analysis of the layers of FR4 composite materials is recommended.

The dynamic response of the PCB depends on the characteristic parameters of the materials. A PCB structure is often composed of four layers of copper and three layers of an FR4 composite material. This FR4 material is a combination of fiberglass-type reinforcement and an epoxy resin. The FR4 laminate is a stack of a number Np of fiber preform layers characterized by a thickness hi, an orientation angle θ_i and a volume ratio of fibers V_f. The optimization of the PCB design consists of reducing weight and ensuring dynamic performance. An optimization procedure is usually developed to determine the optimal values of the PCB parameters subject to the frequency range constraint.

There are three main steps in the optimization process of a PCB (Figure 4.20). The first step consists of incorporating the structural variables characterizing the FR4 composite material in a digital finite element model. The second step involves the effect of the volume ratio of the fibers and the orientation of the folds on the vibration response. In the last step, design is optimized to reduce mass and improve mechanical performance.

The PCB under study is a rectangular plate with the following dimensions: 170 mm x 130 mm x 1.6 mm (Figure 4.21).

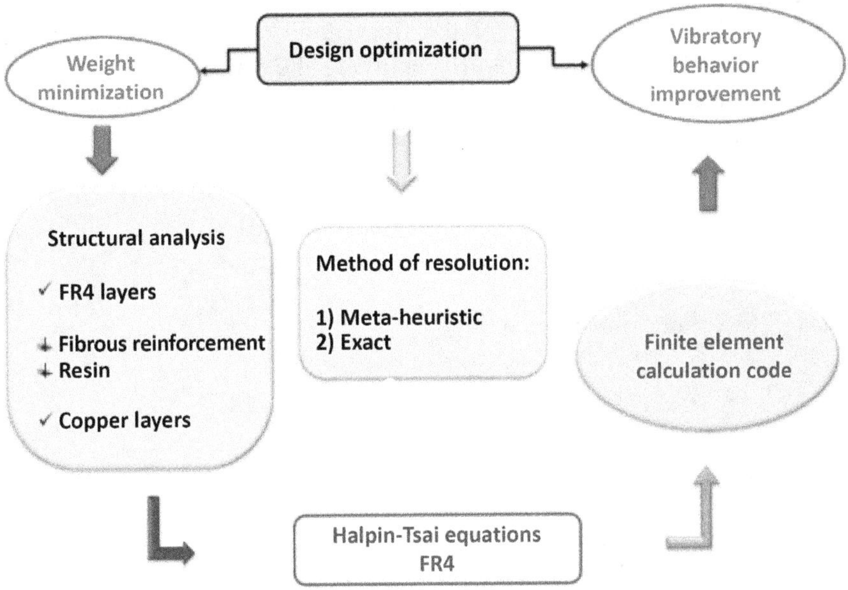

Figure 4.20. *Main steps of the optimization process of a PCB. For a color version of this figure, see www.iste.co.uk/dahoo/metrology2.zip*

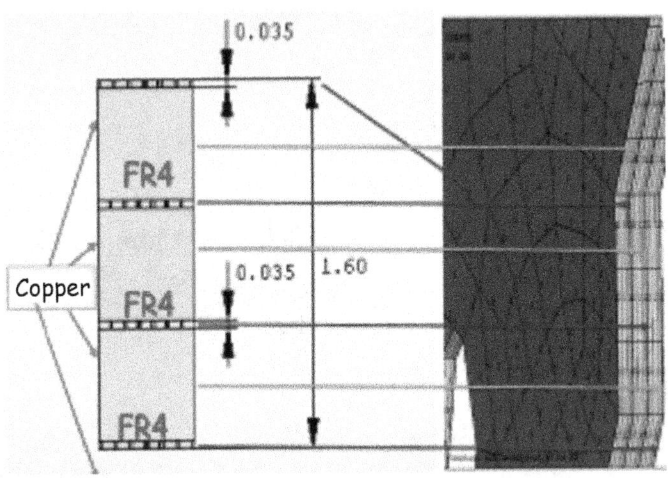

Figure 4.21. *PCB cross-section and finite element mesh. For a color version of this figure, see www.iste.co.uk/dahoo/metrology2.zip*

The elastic moduli of FR4 composite materials can be expressed as a function of the elastic moduli of the fiber reinforcement and the matrix, using the Halpin–Tsai equations. By applying the law of mixtures for the moduli (Young's moduli, Poisson's coefficients), the FR4 Young moduli and Poisson coefficients can be expressed as:

$$\begin{aligned} E_{x,y,Fr4} &= E_{x,y,f} V_f + E_m(1-V_f) \\ \nu_{xy,Fr4} &= \nu_{xy,f} V_f + \nu_m(1-V_f) \end{aligned} \qquad [4.13]$$

The FR4 volume density is written as:

$$\rho_{Fr4} = \rho_f V_f + \rho_m(1-V_f) \qquad [4.14]$$

The other moduli are obtained from:

$$\frac{M_{Fr4}}{M_m} = \frac{1+\xi\eta V_f}{1-\eta V_f} \qquad [4.15]$$

where η is expressed by:

$$\eta = \frac{\left(\frac{M_f}{M_m}\right) - 1}{\left(\frac{M_f}{M_m}\right) + \xi} \qquad [4.16]$$

where:

– f is the index relating to fibers;

– m is the index relating to the matrix;

– M is the transverse Young's modulus of shear or the transverse Poisson's ratio;

– M_f is the corresponding modulus of fibers;

– M_m is the module corresponding to the matrix.

The ξ factor is a characteristic parameter of the fiber reinforcement, which depends on the geometry of the fibers, the arrangement of the fibers and the modulus considered.

Here, $\xi = 2$ is used to determine the Young modulus, while $\xi = 1$ is used to determine the shear modulus.

4.5.4.2. *Optimization and uncertainties*

The effect of the volume ratio of fibers V_f and the stacking sequences of layers θ on the natural frequencies characterizing the modes of vibration of the PCB structure is studied. The values of the parameters characterizing the copper, the fiber and the matrix are given in Table 4.11. In order to measure the impact of the design variables such as the orientation and the volume ratio of the fibers on the eigenmodes, the orientation of the folds is assumed to be identical $\theta_i = \theta$.

Parameter/unit	Value
ρ_{cu} (kg/m^3)	8930
ρ_f (kg/m^3)	2750
ρ_m (kg/m^3)	1200
V_f (%)	10
E_{xf} (Gpa)	72.5
E_{yf} (Gpa)	72.5
G_{xyf} (Gpa)	30
E_m (Gpa)	2.6
G_m (Gpa)	0.985
v_f	0.2
v_m	0.32

Table 4.11. *Values of the parameters used for numerical simulations*

Figure 4.22 shows the results obtained as iso-surfaces. For the first mode, where the natural frequency of the printed circuit must be greater than 169 Hz, the optimal choice is bounded by the range intervals defined by $42° \leq \theta \leq 60°$ and $35\% \leq V_f \leq 40\%$. For the second mode, where the maximum assigned frequency is greater than 216 Hz (Figure 4.22), the optimal choice is bounded by the range intervals defined by $0° \leq \theta \leq 22°$ and $35\% \leq V_f \leq 40\%$, $78° \leq \theta \leq 90°$ and $35\% \leq V_f \leq 40\%$. Such a parametric study makes it possible to define for each vibrational mode the feasible domain of the optimal design subject to the frequency constraint.

The objective of setting the architecture parameters of the fiber preform is to improve the mechanical performance of the PCB structure. This consists of choosing the values for the volume ratio of the fibers, the orientation and the thickness of both the plies and the copper layer. The objective function of this multi-constraint optimization problem is expressed as:

$$\begin{cases} F_{obj}(X) = \dfrac{1}{\rho_{Cu}\, h_{Cu} + \rho_{FR4}(V_f)\, h_{FR4}} \\ h_{FR4} = N_{pl} h_{pl} \\ X = \left(V_f, h_{FR4}, \theta, h_{Cu}\right) \end{cases} \quad [4.17]$$

where F_{obj} is a nonlinear function of the design variables grouped in the vector X, h_{pl} is the thickness of the ply, h_{FR4} is the thickness of the FR4 for a number of plies N_{pl} and h_{Cu} is the thickness of the copper layer. The choice of the effective parameters of the PCB greatly contributes to the determination of F_{obj}. The number of design variables depends on the properties of the materials (copper, fibers and resin).

Minimizing the mass of the PCB is equivalent to maximizing the objective function F_{obj}. The constraints imposed on the limits of the resolution space are given by:

$$\begin{cases} 0.1 \le V_f \le 0.4 \\ h_{FR4\_m} \le h_{FR4} \le h_{FR4\_M} \\ h_{Cu\_m} \le h_{Cu} \le h_{Cu\_M} \\ 0 \le \theta \le 90 \end{cases} \quad [4.18]$$

where h_{FR4_m} and h_{FR4_M} are, respectively, the lower and upper bound of the range of variation of h_{FR4}, and h_{Cu_m} and h_{Cu_M} are, respectively, the lower and upper bound of the variation interval of h_{FR4}. The frequency constraints imposed on the *i*th mode of vibration are expressed by the following inequalities:

$$\forall i \in IN \qquad f_i(X) \ge f_{i,b} \quad [4.19]$$

where f_i and $f_{i,b}$ represent, respectively, the natural frequency and the lower frequency limit assigned to each *i*th mode.

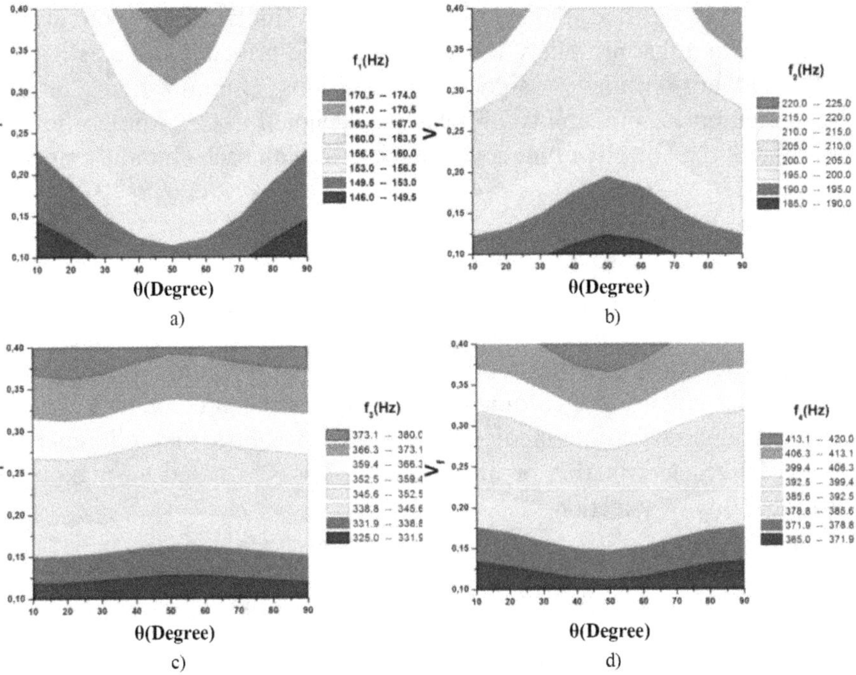

Figure 4.22. *Iso-surfaces of the effects of the volume ratio and the fiber orientation on the natural frequencies f_1 (a), f_2 (b), f_3 (c) and f_4 (d) of a PCB. For a color version of this figure, see www.iste.co.uk/dahoo/metrology2.zip*

As it is impossible to determine an analytic closed-form solution to calculate an optimum value of the objective function, a meta-heuristic solving method based on a genetic algorithm, which mimics the mechanisms of natural selection and genetics, is applied. This algorithm uses the principles of survival of the best-suited individuals. It is modeled on the natural evolution of species. In this context, the various processes observed in genetics (selection, crossing, mutation, etc.) are implemented.

A population P_0 containing N_p chromosomes (solutions) is generated by a random process (Figure 4.20). In order to create successive generations, the chromosomes must undergo a correction process to verify the constraints imposed on the optimization problem. This requires applying the FE (finite elements) code for each generation of solution until the frequency constraints are verified. Selection helps identify individuals belonging to a population and that are likely to be crossed. The principle of selection uses the Goldberg roulette algorithm. In this optimization problem (maximization), a probability of selection, denoted as $P_{sel,i}$, proportional to the value of the objective function is associated with each chromosome m:

$$\forall i \in \{1,\ldots,N_p\} \qquad P_{sel,i} = \frac{f(X_i)}{\sum_{i=1}^{N_p} f(X_i)} \qquad [4.20]$$

Each chromosome is thus reproduced with the probability $P_{sel,i}$. Then, the solutions are classified according to three classes: the good ones (C_b) are the most reproduced, the intermediate (C_i) and the weak (C_f) are eliminated. In practice, the determination of the different classes is based on maximizing the probability of selection:

$$\forall i \in \{1,\ldots,N_p\} \qquad \begin{cases} Chr_i \in C_b \Rightarrow P_{sel,i} \geq \lambda_s P_{s,\max} \\ Chr_i \in C_i \Rightarrow 0.5\lambda_s P_{s,\max} \leq P_{sel,i} < \lambda_s P_{s,\max} & /0 < \lambda_s < 1 \\ Chr_i \in C_f \Rightarrow P_{sel,i} < 0.5\lambda_s P_{s,\max} \end{cases} \qquad [4.21]$$

This probabilistic aspect is achieved by crossing the classes determined during the selection process. The oldest crossover operators used are the one-dot and two-dot operator on two binary-coded chromosomes. The mutation operator provides genetic algorithms with the necessary randomness for efficient exploration of the hyperspace under study. This operator ensures that the genetic algorithm achieves most of the achievable points. In fact, the genes of the chromosomes chosen randomly are mutated with a low probability of P_m mutation. The crossing process increases the diversity of the population by manipulating the components of chromosomes. Conventionally, the crossing starts with two parents and generates two children; it is applied with a probability.

4.5.4.3. *Analysis of simulation results*

The frequency constraint is imposed on the first mode with a value $f_{1,b}$. The population contains 50 chromosomes and the total number of iterations is 125. Table 4.12 provides the values of the parameters used in the numerical simulation.

Numerical parameter	Value
Np	50
λs	0.7
Iter	125
Pm	0.06
$f_{l,b}$	200

Table 4.12. *Values of the parameters used in the genetic algorithm*

Figures 4.23 and 4.24 show the transition history of the genetic algorithm process, which converges to the optimal variable values of the parameters of the fibrous reinforcements and the copper layer. Convergence is very slow for the volume ratio of the fibers (107 iterations). It is faster for both the ply orientation angle and the ratio of the thicknesses of FR4 relative to the copper layer (Figure 4.24). The optimal values of the optimization variables are summarized in Table 4.13.

Variables	V_f (%)	$h_{FR4}(mm)$	θ (°)	$h_{Cu}(mm)$
Values	10.38	0.7542	20.0135	0.03736

Table 4.13. *Optimal values of PCB design variables*

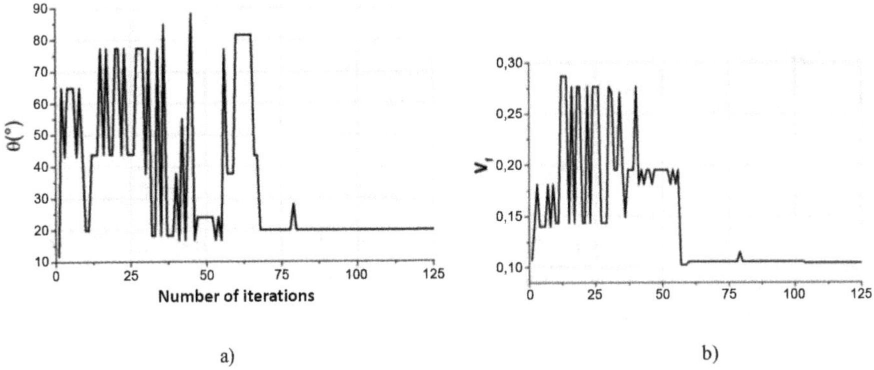

Figure 4.23. *Evolution of the orientation angle (a) and the volume ratio of fibers (b) according to the iterations*

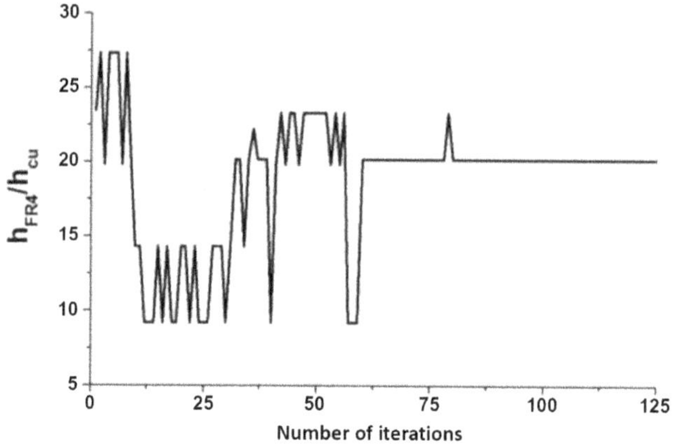

Figure 4.24. *Evolution of the ratio between the thickness of FR4 and the copper layer according to the iterations*

4.6. Reliability-based design optimization in nanotechnology

In [DAH 16] and [TAN 15], a description of carbon nanotubes and the main parameters on which their properties are based is given.

The elastoplastic properties of thin-film single-walled carbon nanotube (SWCNT) structures are estimated by combining the finite element method results with the data from nanoindentation tests. The FEM provides the limits where the test data are expected. The RBDO method is applied to the load–displacement curve distribution to analyze the reliability of the estimate.

4.6.1. *Thin-film SWCNT structures*

Based on the full charge–discharge cycles of the indentation of thin-film SWCNT structures, the elastoplastic behavior of the SWCNT film is described by a linear expression.

To simulate the stress–strain behavior, the following bilinear model is used:

$$\sigma = \begin{cases} E\varepsilon, & for\ \sigma \leq \sigma_Y \\ \sigma_Y + E_t(\varepsilon - \varepsilon_Y), & for\ \sigma \geq \sigma_Y \end{cases} \quad [4.22]$$

where σ_Y and ε_Y are the yield strength and the strain, respectively, and $\varepsilon_Y = \sigma_Y/E$, E is the Young modulus, and E_t is the shear modulus.

Based on the linear elasto-plastic model, the reduced modulus E_r during unloading is modified (Figure 4.25). The modulus of phase 2 is expressed by analogy as:

$$E_r^* = \frac{1}{2}\frac{\sqrt{\pi}}{\sqrt{A}}\frac{dP}{dh}|_{\text{phase2}} \quad [4.23]$$

where E_r^* is the reduced modulus of phase 2, and $\frac{dP}{dh}|_{\text{phase2}}$ is the slope of the last part of the unloading curve.

According to contact mechanics, the tangent modulus E_t is obtained by:

$$\frac{1}{E_r^*} = \frac{(1-\nu^2)}{E_t} + \frac{(1-\nu_i^2)}{E_i} \quad [4.24]$$

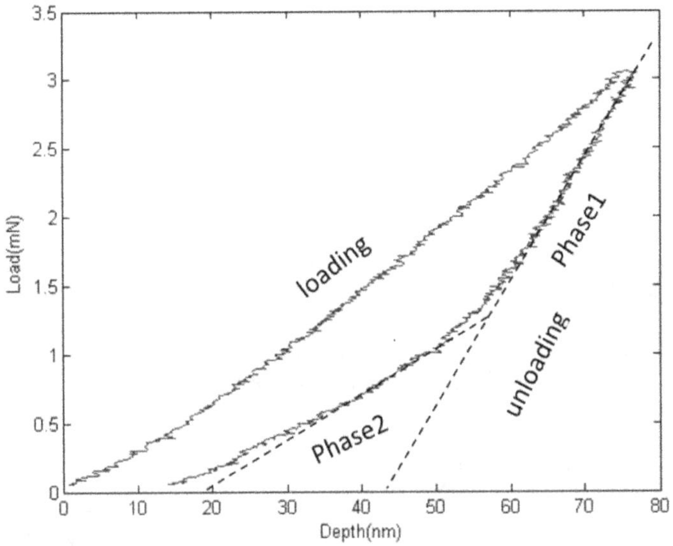

Figure 4.25. *Load exerted on a thin-film SWCNT structure as a function of the displacement of the indenter during a nanoindentation measurement. For a color version of this figure, see www.iste.co.uk/dahoo/metrology2.zip*

Due to the complexity of the phase change, the charge–displacement (p-h) relationship of a thin-film SWCNT structure during indentation is implicitly written as:

$$P = P(h, E_r, \sigma_y, E_t, \theta) \quad [4.25]$$

where θ is the angle at the apex of the indenter.

The material property assessment program uses the analysis of uncertainties based on the finite element calculation. The load–displacement curve is obtained by simulation:

$$\textbf{min}: \ \left\| \boldsymbol{p}^i - \overline{\boldsymbol{p}}_t \right\| \quad [4.26]$$

Under

$$\left| \frac{\overline{h}_{max} - h^i_{max}}{\overline{h}_{max}} \right| \leq \Delta_1 \quad [4.27]$$

$$\left| \frac{\overline{s} - s^i}{\overline{s}} \right| \leq \Delta_2 \quad [4.28]$$

where:

– $p^i p^i p^i$ is the vector of the charge of the ith iteration, and $\overline{p}_t$ is the vector of the average charge in the indentation test;

– $\overline{h}_{max}$ and $\bar{s}$ are the average values of the maximum displacement and the stiffness of contact, respectively;

– $\Delta_1\Delta_1\Delta_2\Delta_2\Delta_1$ and Δ_2 are the limits of variation of the maximum displacement (h_{max}) and the stiffness of the contact (S).

4.6.2. *Digital model of thin-film SWCNT structures*

4.6.2.1. *The initial properties of the materials used for the measurements*

For the indenter, the Young modulus is 1.143 GPa and the Poisson ratio is 0.07. For the silicon substrate, the Young modulus is 180 GPa and the Poisson ratio is 0.278 [TAN 15]. The Young modulus of the thin-film

SWCNT is 192.83 ± 13.922 GPa; the initial value of the elastic limit Y_0 is 4.2 GPa and the Poisson ratio ν is 0.18, according to [TAN 15].

4.6.2.2. *Construction of the finite element model*

The model of the material response uses the criterion of "von Mises with bilinear isotropic hardening". The indenter, the thin-film SWCNT structure and the substrate are meshed by 20-node 3D solid elements. The interaction between the indenter and the specimen is modeled as a frictionless surface-to-surface contact. The interface between the film and the substrate is assumed to be bonded. The meshes that are close to the indenter are refined to describe the strain and stress gradient with precision (Figure 4.26). A medium force is continuously applied to the top surface of the indenter in the z-direction. All the degrees of freedom of the lower nodes of the substrate are fixed. The predetermined maximum value of the force is 3 mN.

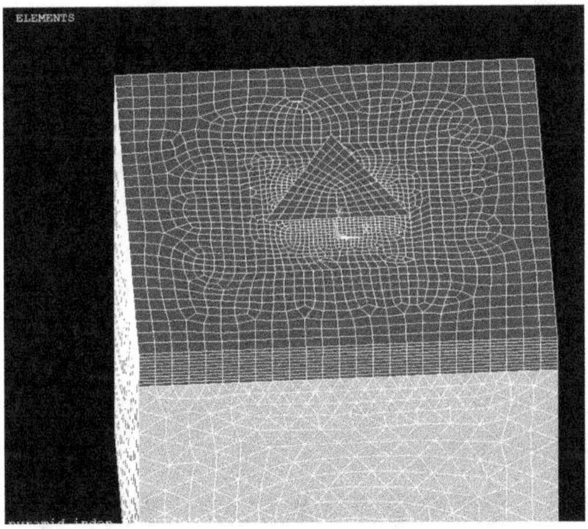

Figure 4.26. *Finite element model of the indenter–film system. For a color version of this figure, see www.iste.co.uk/dahoo/metrology2.zip*

4.6.3. *Numerical results*

Figure 4.27 shows the comparison between the experimental and simulation results.

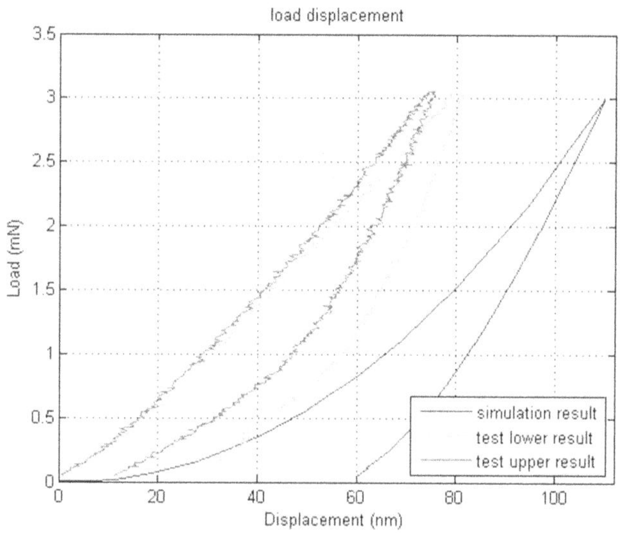

Figure 4.27. *Results of testing and modeling the load of a thin-film SWCNT structure as a function of displacement. For a color version of this figure, see www.iste.co.uk/dahoo/metrology2.zip*

The dispersion of the results is caused by the approximations of the model, the uncertainties of the properties of the material tested and the quality of the contact surface of the indenter.

Figure 4.28 shows the strain and stress distributions of the film–substrate SWCNT system.

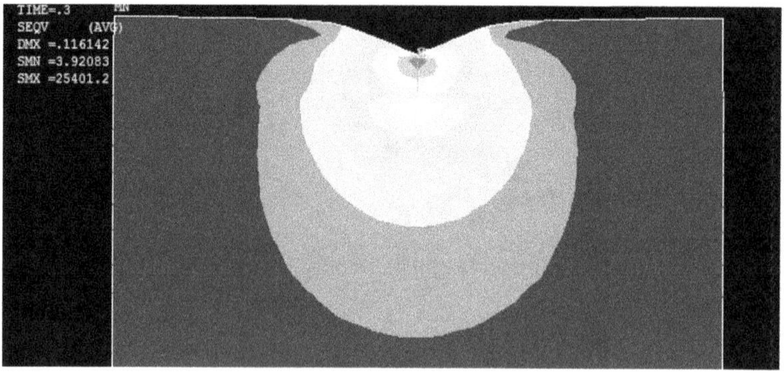

Figure 4.28. *Stress distribution of the film–substrate SWCNT system. For a color version of this figure, see www.iste.co.uk/dahoo/metrology2.zip*

Parametres		θ°C	σ_y (Gpa)	E_t (Gpa)		$h_{\max}$		$\frac{dp}{dh}$	
						$h_{\max}$ mn	Δ_1	$\frac{dp}{dh}$ mN/nm	Δ_2
Average experimental value						77.68		0.0963	
FE Simulation	Iteration 1	65.3	4.2	42		108.881	40.17%	0.0867	9.97%
	Iteration 2	70	4.2	42		87.59	12.76%	0.10845	12.62%
	Iteration 3	70	2.1	42		90.4407	16.43%	0.114	18.38%
	Iteration 4	70	8.385	21		85.903	10.6%	0.10256	6.5%
	Iteration 5	70	8.385	31.5		84.6446	8.97%	0.10098	4.86%
	Iteration 6	70	8.385	42		82.8116	6.61%	0.09858	2.37%

Table 4.14. *Results of finite element simulations*

Figure 4.29 compares the load–displacement characteristics of thin-film SWCNT structures for various indenter shapes and the same maximum load.

A shape imperfection of the indenter directly affects the contact area. The shape errors of the indenter in finite element simulation are compensated by modifying the angle of the vertex of the indenter by approximations:

$$A_{proj} = c_2 {h_c}^2 + c_1 h_c + c_0 = 3\sqrt{3}{h_c}^2 tan^2\theta \qquad [4.29]$$

Table 4.14 summarizes the results of the iterations of the simulations. After five iterations, the simulation results are close to the experimental data (Figures 4.30, 4.31, 4.32 and 4.33).

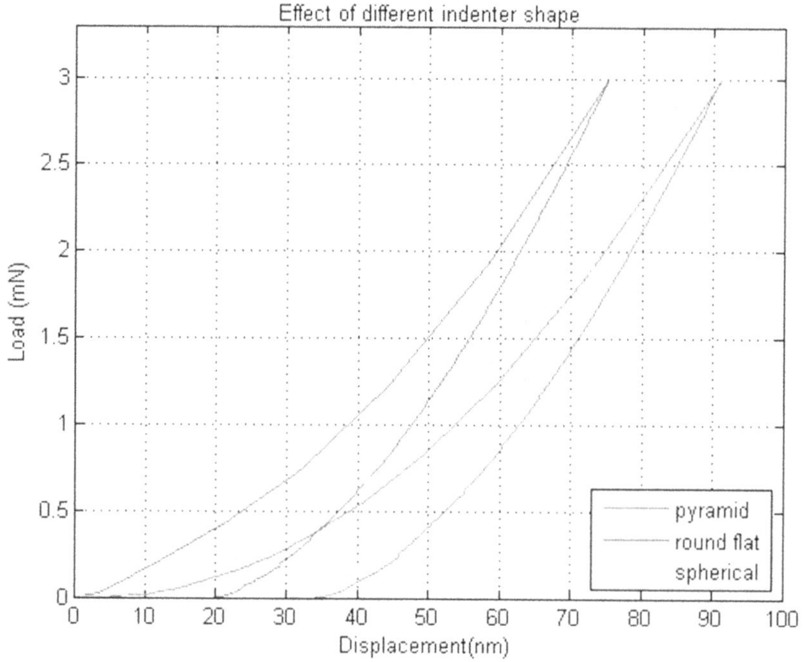

Figure 4.29. *Effects of different forms of indenter on the load–displacement curve. For a color version of this figure, see www.iste.co.uk/dahoo/metrology2.zip*

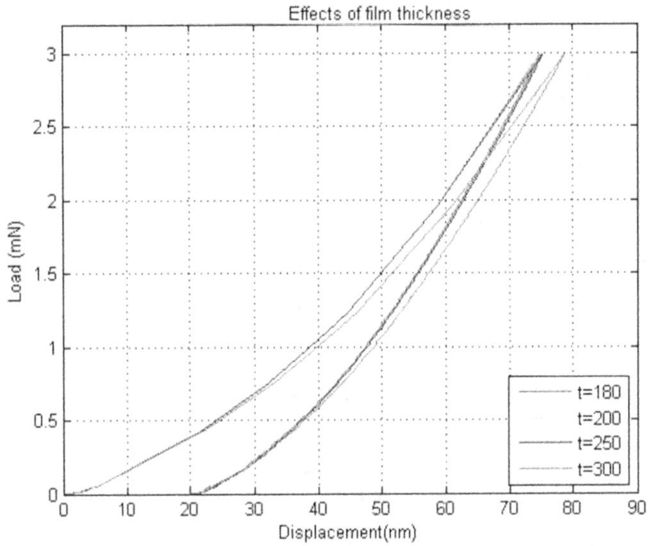

Figure 4.30. *Effect of the thickness of the SWCNT structure on the load–displacement curve. For a color version of this figure, see www.iste.co.uk/dahoo/metrology2.zip*

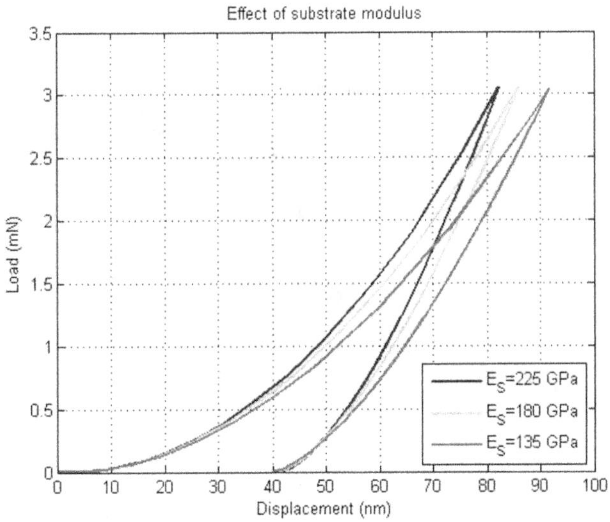

Figure 4.31. *Effect of the Young modulus of the substrate on the load–displacement curve. For a color version of this figure, see www.iste.co.uk/dahoo/metrology2.zip*

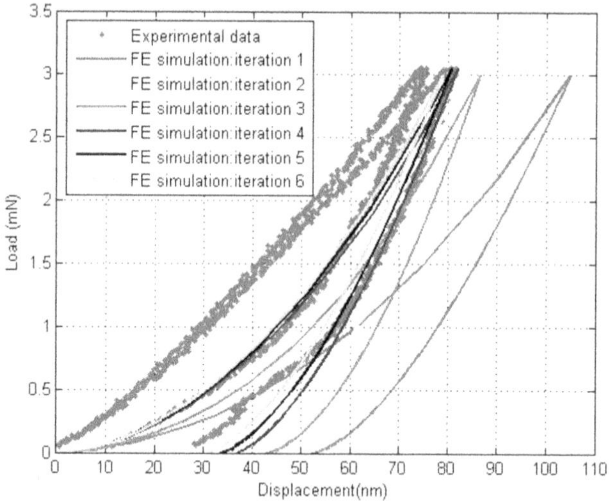

Figure 4.32. *Load–displacement curves from testing and modeling. For a color version of this figure, see www.iste.co.uk/dahoo/metrology2.zip*

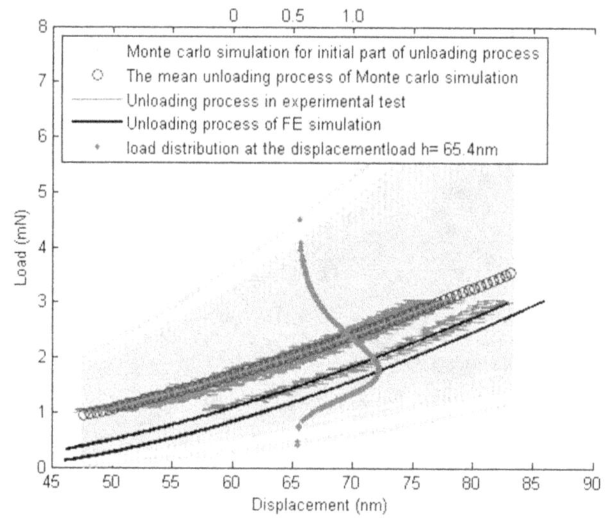

Figure 4.33. *Experimental and simulated discharge curves. For a color version of this figure, see www.iste.co.uk/dahoo/metrology2.zip*

Figure 4.33 shows the distribution of unloading data from the experiment, Monte Carlo simulation and finite element simulations. The numerical results are close to the experimental data.

4.7. Conclusion

In this chapter, the deterministic design optimization (DDO) and the reliability-based design optimization (RBDO) methods were presented. These methods can be applied in the initial design phases of mechanical structures. Compared to the DDO method, the RBDO method leads to more efficient designs.

The RBDO method consists of solving a mechanical structure design problem in a single hybrid space combining two types of variables: the design variables and the random variables. This hybrid space is used to set the various parameter values of the problem. Several applications of RBDO were given, in particular at the nanometer scale for the carbon nanotube.

Appendix

Short Overview of Quantum Mechanics

In quantum mechanics (QM), the Hilbert space (Ω) of the states of a system of several particles is the tensor product of the space states (Ω_k) of each particle ($\Omega = \Omega_1 \otimes \Omega_2 \otimes \Omega_3 \ \ldots \ \otimes \Omega_k \ldots$). As the kets should have a physical meaning, this is not the case if the particles are identical. The physical kets are either symmetrical ($\Omega_S \subset \Omega$, integer spin bosons, photon, meson, etc.) or antisymmetrical ($\Omega_A \subset \Omega$, half-integer spin fermions, electron, positron, muon, etc.) in a permutation of the particles. Consequently, only certain kets of the Hilbert space of identical particles can describe their physical states.

In QM, the classical variables, position vector and momentum are described by Hermitian operators that verify the commutation rules $[x,p] = i\hbar$ that operate on a space of quantum states that are functions whose squares are integrable (or kets $|\Psi\rangle$ of a Hilbert space). The formalism of ket $|\Psi\rangle$ and bra $\langle\Psi|$, which is a notation independent of the representation (x or p), was introduced by P.A.M. Dirac to simplify the QM notation. Hilbert space states are determined by the stationary Schrödinger eigenvalue equation:

$$H|\psi_k\rangle = E_k|\psi_k\rangle \qquad \text{[A.1]}$$

where H is the Hamiltonian operator of the physical system considered, which represents the sum of its kinetic energy T and its potential energy V. The eigenfunctions are mutually orthogonal. They are generally standardized ($\langle\psi_i|\psi_k\rangle = \delta_{ik}$) and define a complete set. Each state vector $|\Psi\rangle$ of the physical system is expressed as a linear combination of $|\psi_k\rangle$, for example $|\Psi\rangle = \sum_{k=0}^{\infty} C_k |\psi_k\rangle$. As the C_k are determined by $C_k = \langle\psi_k|\Psi\rangle$, the state vector can be written as $|\Psi\rangle = \sum_{k=0}^{\infty} |\psi_k\rangle\langle\psi_k|\Psi\rangle$, where $P_k = |\psi_k\rangle\langle\psi_k|$ is the projection operator. This operator verifies the relation $P_k^2 = P_k$. The Hilbert space of a system Σ composed of several systems Σ_i, such that $\Sigma = U_i\Sigma_i$ (countable set), is the tensor product of the Hilbert spaces of each subsystem such that $H_\Sigma = \prod_{\otimes i} H_{\Sigma_i}$. For example, the Hilbert space associated with the degrees of freedom of vibration, rotation and spin of a molecule when it is in a fundamental electronic state is written as $H_{VRS} = H_V \otimes H_R \otimes H_S$ at zero order.

In QM, a quantum system can be prepared in a state described by a vector $|\Psi\rangle$ of the Hilbert space. In theory, if all possible states $|\Psi\rangle$ are determined and known, so are the probabilities of all possible results of a given measurement. Such states are called pure states, and the probability attached to each measurement is given by the principles of QM.

The time evolution of the system is determined by the time-dependent equation: $\hat{H}|\psi\rangle = i\hbar \frac{\partial}{\partial t}|\psi\rangle$. If $\hat{H}$ does not depend on time, then $|\psi\rangle(t) = e^{-i\hat{H}t/\hbar}|\psi\rangle(0)$. The latter equation can be written as $|\psi\rangle(t) = \hat{U}(t)|\psi\rangle(0)$, where $\hat{U}(t)$ is the temporal evolution operator of the physical system such that $\hat{U}(t) = e^{-i\hat{H}t/\hbar}$. When the knowledge of the state of the system is incomplete, the description of the quantum system requires the introduction of a density operator.

In QM, there are two postulates associated with the result of a measurement on an observable:

– Postulate 1: "a physical observable is represented by a hermetic operator A and the result of its measurement is one of the eigenvalues of the operator with a certain probability".

– Postulate 2: "if the quantum system is in a state $|\Psi\rangle$, the average value of an observable is given by: $\langle\Psi|A|\Psi\rangle$". If p_k is the probability that the state

$|\psi_k\rangle$ is known, the mean that takes into account the quantum and statistical aspect has the expression: $\langle \hat{A} \rangle = \sum_k p_k \langle \psi_k | \hat{A} | \psi_k \rangle = Tr(\hat{\rho}\hat{A})$, where Tr represents the trace (sum of the diagonal elements) of the matrix $\hat{\rho}\hat{A}$. Here, $\hat{\rho} = \sum_k p_k |\psi_k\rangle\langle\psi_k|$ is the density matrix of the system, where $|\psi_k\rangle\langle\psi_k| = P_k$ is the projection operator.

The density matrix is a description of a quantum system that is composed of a statistical mixture of states. The mean value of an operator associated with an observable includes a quantum mean and a classical statistical mean. These mean values cannot be separated in the density matrix.

A pure state is represented by an eigenvector of a Hermitian operator of a Hilbert space, but a mixed state is a statistical state represented by a density matrix that is not a vector of a Hilbert space. In this case, the time evolution of the density matrix is given by the time-dependent von Neumann equation:

$$i\hbar \frac{\partial}{\partial t} \hat{\rho} = \left[\hat{H}, \hat{\rho} \right] = \hat{H}\hat{\rho} - \hat{\rho}\hat{H} \quad \text{[A.2]}$$

Just as the space of quantum states of a physical system, functions whose squares are integrable (or kets $|\psi\rangle$), is a Hilbert space, the space of operators acting on quantum states constitutes the set of linear transformations $L(H_\Sigma)$ acting on the Hilbert space of quantum states. The construction of a complete basis of $L(H_\Sigma)$ can be carried out as a result of the properties of the tensor product in a vector space of infinite dimension and the fact that any operator of $L(H_\Sigma)$ is a representation of the group O(3).

References

[ABB 73] ABBE E., "Beiträge zur theorie des mikroskops und der mikroskopischen wahrnehmung", *Arch. Mikroskop. Anat.* vol. 9, pp. 413–418, 1873.

[ARU 19] ARUTE F., ARYA K., BABBUSH R. *et al.*, "Quantum supremacy using a programmable superconducting processor", *Nature*, vol. 574, no. 7779, pp. 505–510, 2019.

[ASH 72] ASH E.A., NICHOLLS G., "Super-resolution aperture scanning microscope", *Nature*, vol. 237, p. 510A, 1972.

[BAR 48] BARDEEN J., BRATTAIN W.H., "The transistor, a semi-conductor triode", *Phys. Rev.*, no. 74, pp. 230–231, 1948.

[BET 86] BETZIG E., LEWIS A., HAROOTUNIAN A. *et al.*, "Near-field scanning optical microscopy (NSOM): Development and biophysical applications", *Biophys. J.*, vol. 49, pp. 269–279, 1986.

[BET 87] BETZIG E., ISAACSON M., LEWIS A., "Collection mode near-field optical microscopy", *Appl. Phys. Lett.*, vol. 51, pp. 2088–2090, 1987.

[BIN 82a] BINNIG G., ROHER H., GERBER C. *et al.*, "Tunneling through a controllable vacuum gap", *Appl. Phys. Lett.*, vol. 40, pp. 178–180, 1982.

[BIN 82b] BINNIG G., ROHER H., "Scanning tunneling microscopy", *Helv. Phys. Acta*, vol. 55, pp. 726–735, 1982.

[BIN 83] Binnig G., ROHER H., "Scanning tunneling microscopy", *Surf. Sci.*, vol. 126, pp. 236–244, 1983.

[BIN 86a] BINNIG G., ROHRER H., "Scanning tunneling microscopy", *IBM J. Res. Dev.*, vol. 30, pp. 355–369, 1986.

[BIN 86b] BINNIG G., QUATE C.F., GERBER C., "Atomic force microscope", *Phys. Rev. Lett.*, vol. 56, pp. 930–933, 1986.

[BIO 56] BIOT M.A., "Thermoelasticity and irreversible thermodynamics", *J. Appl. Phys.*, vol. 27, pp. 240–253, 1956.

[BRI 38] BRILLOUIN L., *Les tenseurs en mecanique et en élasticité*, Editions Masson et Cie, Paris, 1938.

[BRU 55] BRUHAT G., *Mécanique*, Editions Masson et Cie, Paris, 1955.

[BRU 62] BRUHAT G., *Thermodynamique*, Editions Masson et Cie, Paris, 1962.

[CAR 16] CARDON A., ITMI M., *New Autonomous Systems*, ISTE Ltd, London and John Wiley & Sons, New York, 2016.

[CHU 98] CHUANG I.L., GERSHENFELD N., KUBINEC M., "Experimental implementation of fast quantum searching", *Phys. Rev. Lett.*, vol. 80, no. 15, p. 3408, 1998.

[COA 12] COANGA J.M., ALAYLI N., DAHOO P.R., "Metrological applications of ellipsometry with temperature change", *Proceedings of 4th International Metrology Conference 23–27 April 2012 (CAFMET 2012) Marrakech, Morocco*, vol. 1, Curran Associates, Inc., July 2012.

[COL 72] COLELLA R., MENADUE J.F., "Comparison of experimental and beam calculated intensities for glancing incidence high-energy electron diffraction", *Acta. Cryst.*, vol. A28, pp. 16–22, 1972.

[COU 89] COURJON D., SARAYEDDINE K., SPAJER M., "Scanning tunneling optical microscopy", *Opt. Commun.*, vol. 71, pp. 23–281, 1989.

[CUR 80] CURIE J., CURIE P., "Développement, par pression, de l'électricité polaire dans les cristaux hémièdres à faces inclinées", *Comptes Rendus de l'Académie des Sciences*, vol. XCI, pp. 294–295, 2 August 1980.

[CUR 81] CURIE P., CURIE J., "Contractions et dilatations produites par des tensions électriques dans les cristaux hémièdres à faces inclinées", *Comptes Rendus de l'Académie des Sciences*, vol. XCIII, pp. 1137–1140, 26 December 1981.

[DAH 99] DAHOO P.R., BERRODIER I., RADUCU V. *et al.*, "Splitting of ν_2 vibrational mode of CO_2 isotopic species in the unstable trapping site in argon matrix", *Eur. Phys. J. D*, vol. 5, pp. 71–81, 1999.

[DAH 03] DAHOO P.R., HAMON T., SCHNEIDER M. *et al.*, "Ellipsometry: Principles, signal processing and applications to metrology", *Proceedings CIMNA*, Lebanon, 2003.

[DAH 04a] DAHOO P.R., GIRARD A., TESSEIR M. *et al.*, "Characterizaton of pulsed laser deposited SmFeO3 morphology: Effect of fluence, substrate temperature and oxygen pressure", *Appl. Phys. A*, vol. 79, pp. 1399–1403, 2004.

[DAH 04b] DAHOO P.R., HAMON T., NEGULESCU B. *et al.*, "Evidence by spectroscopic ellipsometry of optical property change in pulsed laser deposited NiO films when heated in air at Neel temperature", *Appl. Phys. A*, vol. 79, pp. 1439–1443, 2004.

[DAH 06] DAHOO P.R., LAKHLIFI A., CHABBI H. *et al.*, "Matrix effect on triatomic CO_2 molecule: Comparison between krypton and xenon", *J. Mol. Struct.*, vol. 786, pp. 157–167, 2006.

[DAH 11] DAHOO P.R., "Metrological applications of ellipsometry", *Proceedings of the 3rd International Metrology Conference 18–23 April 2010 (CAFMET 2010) Cairo, Egypt*, vol. 1, Curran Associates, Inc., September 2011.

[DAH 16a] DAHOO P.R., POUGNET P., EL HAMI A., *Nanometer-Scale Defect Detection Using Polarized Light*, ISTE Ltd, London, and John Wiley & Sons, New York, 2016.

[DAH 16b] DAHOO P.R., JORGE L., DANIEL C. *et al.*, "Monte Carlo method applied to the ABV model of an interconnect alloy", *Journal of Physics: Conference Series*, vol. 738, pp. 012–073, 2016.

[DAH 17] DAHOO P.R., LAKHLIFI A., *Infrared Spectroscopy of Diatomics for Space Observation 1*, ISTE Ltd, London, and John WIley & Sons, New York, 2017.

[DAH 19] DAHOO P.R., LAKHLIFI A., *Infrared Spectroscopy of Diatomics for Space Observation 2*, ISTE Ltd, London, and John WIley & Sons, New York, 2019.

[DAH 21a] DAHOO P.R., POUGNET P., EL HAMI A., *Applications and Metrology at Nanometer Scale 1*, ISTE, London and JOhn Wiley & Sons, New York, 2021.

[DAH 21b] DAHOO P.R., LAKHLIFI A., *Infrared Spectroscopy of Symmetric and Spherical Top Molecules for Space Observation 1*, ISTE, London, and John Wiley & Sons, New York, 2021.

[DEM 98] DEMAREST F.C., "High-resolution, high-speed, low data age uncertainty, heterodyne displacement measuring interferometer electronics", *Meas. Sci. Technol.*, vol. 9, pp. 1024–1030, 1998.

[DEU 85] DEUTSCH D., "Quantum theory, the Church–Turing principle and the universal quantum computer", *Proc. R. Soc. Lond. A*, vol. 400, no. 1818, pp. 97–117, 1985.

[DEU 92] DEUTSCH D., JOZSA R., "Rapid solution of problems by quantum computation", *Proc. R. Soc. Lond. A*, vol. 439, no. 1907, pp. 553–558, 1992.

[DIJ 87] DIJKAMP D., VENKATESAN T., WU X.D. *et al.*, "Preparation of Y-Ba-Cu oxide superconductor thin films using pulsed laser evaporation from high T_C bulk material", *Appl. Phys. Lett.*, vol. 51, pp. 619–621, 1987.

[DIR 28] DIRAC P.A.M., "The quantum theory of the electron", *Proc. R. Soc. Lond. A*, vol. 117, no. 778, pp. 610–24, 1928.

[DIV 97] DIVINCENZO D.P., "Topics in quantum computers", *Mesoscopic Electron Transport*, Springer, pp. 657–677, 1997.

[DUP 06] DUPAS C., HOUDY P., LAHMANI M., *Nanoscience, Nanotechnologies and Nanophysics*, Springer-Verlag Berlin and Heideberg GmbH & Co Eds, 2006.

[ELH 10] EL HAMI A.C.-A., *Design Optimization in Computational Mechanics*, ISTE, London, and John Wiley & Sons, New York, 2010.

[ELH 11] EL HAMI A., RADI B., *Fiabilité et optimisation des systemes : theorie et applications, cours et exercices corriges*, Ellipses, 2011.

[ELH 13] EL HAMI A., RADI B., *Uncertainty and Optimization in Structural Mechanics*, ISTE, London, and John Wiley & Sons, New York, 2013.

[FIS 88] FISCHER E., DÜRIG U., POHL D., "Near-field scanning optical microscopy in reflection", *Appl. Phys. Lett.*, vol. 52, pp. 249–251, 1988.

[FRA 05] FRANTA D., NEGULESCU B., THOMAS L. *et al.*, "Optical properties of NiO thin films prepared by pulsed laser deposition technique", *Appl. Surf. Sci.*, vol. 244, p. 426, 2005.

[FUW 15] FUWA M., TAKEDA S., ZWIERZ M. *et al.*, "Experimental proof of nonlocal wavefunction collapse for a single particle using homodyne measurements", *Nat. Commun.*, vol. 6, p. 6665, 2015.

[GRO 96] GROVER L.K., "A fast quantum mechanical algorithm for database search", *Proceedings of the Twenty-Eighth Annual ACM Symposium on Theory of Computing*, ACM, pp. 212–219, 1996.

[GRO 01] GROSSHANS F., GRANGIER P., "Effective quantum efficiency in the pulsed homodyne detection of a n-photon state", *Eur. Phys. J. D*, vol. 14, pp. 119–125, 2001.

[HAM 06] HAMON T., BUIL S., POPOVA E. *et al.*, "Investigation of a one dimensional magnetophotonic crystal for the study of ultrathin magnetic layer", *J. Phys. D: Appl. Phys.*, vol. 39, pp. 1–6, 2006.

[HAM 07] HAMON T., Cristaux magneto-photonique unidimensionnels : étude du magnetisme de couches ultraminces d'oxydes, PhD thesis, Paris XI, Orsay, 2007.

[HER 45] HERZBERG G., *Molecular Spectra and Molecular Structure II: Infrared and Raman Spectra of Polyatomic Molecules*, VAN NOSTRAND D. (ed.), vol. 2, 1945.

[KHA 02] KHARMANDA, G., MOHAMED A., LEMAIRE M., "Efficient reliability-based design optimization using a hybrid space with application to finite element analysis", *Struct. Multidisc. Optim.*, vol. 24, pp. 233-245, 2002.

[KHE 14] KHETTAB M., Etude de l'influence du résinage au niveau de L'IML (Insulated Metal Leadframe), dans le packaging de module commutateur de courant mécatronique, PhD thesis, UVSQ, Versailles, 2014.

[KIT 80] KITTEL C., KROEMER H., *Thermal Physics*, W.H. Freeman and Company, New York, 1980.

[KIT 03] KITAEV A.Y., "Fault-tolerant quantum computation by anyons", *Ann. Phys.*, vol. 303, no. 1, pp. 2–30, 2003.

[KNE 20] KNEISSL M., KNORR A., REITZENSTEIN S. *et al.*, *Semiconductor Nanophotonics*, Springer Nature Switzerland Eds, Cham, 2020.

[KOK 10] KOK P., LOVETT B.W., *Introduction to Optical Quantum Information Processing*, Cambridge University Press, Cambridge, 2010.

[LAH 04] LAHMANI M., DUPAS C., HOUDY P., *Les nanosciences 1, nanotechnologies et nanophysqiue*, Editions Belin, Paris, 2004.

[LAH 06] LAHMANI M., BRECHIGNAC C., HOUDY P., *Les nanosciences 2, nanomatériaux et nanochimie*, Editions Belin, Paris, 2006.

[LAH 07] LAHMANI M., BOISSEAU P., HOUDY P., *Les nanosciences 3, nanobiotechnologies et nanobiologie*, Editions Belin, Paris, 2007.

[LAH 10] LAHMANI M., BOISSEAU P., HOUDY P., *Les nanosciences 4, nanotoxicologie et nanoéthique*, Editions Belin, Paris, 2010.

[LAN 67] LANDAU L., LIFCHITZ E., *Théorie de l'élasticité*, URSS, Editions MIR, Moscow, 1967.

[LAP 94] LAPLACE P.S., *Oeuvres complètes de laplace*, volume 10, M DCCC XCIV, Editions académies des sciences, Gauthiers-villars et fils imprimeurs de l'ecole polytechnique, Paris, 1894.

[LEE 20] LEE Y.-C., MOON J.-Y., *Introduction to Bionanotechnology*, Springer, GmbH & Co Editions, Berlin and Heideberg, 2020.

[LEW 83] LEWIS A., ISAACSON M., HAROOTUNIAN A. *et al.*, "Scanning optical spectral microscopy with 500å resolution", *Biophys. J.*, vol. 41, pp. 405a–XXX, 1983.

[LIM 14] LIM J., LEE H., LEE S. *et al.*, "Ultrafast Ramsey interferometry to implement cold atomic qubit gates", *Sci. Rep.*, vol. 4, p. 5867, 2014.

[LIN 98] LINDEN N., BARJAT H., FREEMAN R., "An implementation of the Deutsch-Jozsa algorithm on a three-qubit NMR quantum computer", *Chem. Phys. Lett.*, vol. 296, no. 12, pp. 61–67, 1998.

[LIP 81a] LIPPMANN G., "Principe de la conservation de l'électricité", *Annales de chimie et de physique*, vol. 24, pp. 145–178, 1881.

[LIP 81b] LIPPMANN G., "Principe de la conservation de l'électricité ou second principe de la théorie des phénomènes élecriques", *J. Phys. Theor. Appl.*, vol. 10, pp. 381–394, 1881.

[LOS 98] LOSS D., DIVINCENZO D.P., "Quantum computation with quantum dots", *Phys. Rev. A*, vol. 57, no. 1, p. 120, 1998.

[LOU 16] LOURTIOZ J., LAHMANI M., DUPAS-HAEBERLIN C. *et al.*, *Nanosciences and Nanotechnology: Evolution or Revolution?* Springer International Publishing Editions, p. 438, 2016.

[MIC 87] MICHELSON A.A., MORLEY E.W. "On the relative motion of the earth and the luminiferous ether", *The American Journal of Science*, vol. XXXIV, pp. 333–345, 1887.

[MOO 65] MOORE G.E., "Cramming more components onto integrated circuit", *Electronics*, vol. 38, no. 8, 1965.

[NIE 10] NIELSEN M.A., CHUANG I.L., *Quantum Computation and Quantum Information*, Cambridge University Press, Cambridge, 2010.

[NOU 07] NOUN W., BERINI B., DUMONT Y. *et al.*, "Correlation between electrical and ellipsometric properties on high-quality epitaxial thin films of the conductive oxide LaNiO3 on STO (001)", *J. Appl. Phys.*, vol. 102, pp. 063709/1–063709/7, 2007.

[NYE 61] NYE J.F., BLANC D., PUJOL T., *Propriétés physiques des cristaux, leur representation par des tenseurs et des matrices*, Editions Dunod, Paris, 1961.

[OLD 93] OLDHAM N.M., KRAMAR A., HETRICK P.S. *et al.*, "Electronic limitations in phase meters for heterodyne interferometry", *Precision Engineering*, vol. 15, pp. 173–179, 1993.

[POH 84] POHL D.W., DENK W., LANZ M., "Optical stethoscopy: Image recording with resolution λ/20", *Appl. Phys. Lett.*, vol. 44, pp. 651–653, 1984.

[POH 86] POHL D.W., Optical near field scanning microscope, US Patent, 4,604,520, 1986.

[POH 87] POHL D.W., Optical near field scanning microscope, EP patent 0112401, 22 May 1987.

[POU 20] POUGNET P., DAHOO P.R., ALVAREZ J.P., "Highly accelerated testing", in POUGNET P., EL HAMI A. (eds), *Embedded Mechatronic Systems 2*, Revised and Updated 2nd Edition, ISTE Press, London, and Elsevier, Oxford, 2020.

[PRI 62] PRIGOGINE I, *Introduction to Thermodynamics of Irreversible Processes*, John Wiley and Sons Inc, New York, 1962.

[RAT 03] RATNER M., RATNER D., *Nanotechnology: A Gentle Introduction to the Next Big Idea*, Prentice Hall Editions, Upper Saddle River, New Jersey, 2003.

[RED 89] REDDICK R.C., WARMACK R.J., FERREL T.L., "New form of scanning optical microscopy", *Phys. Rev. B*, vol. 39, pp. 767–770, 1989.

[ROJ 08] ROJAS J.E., Méthodologie d'analyse de fiabilité basée sur des techniques heuristiques d'optimisation et modèles sans maillage : applications aux systèmes mécaniques, INSA, Rouen, 2008.

[RUS 33] RUSKA E., "Imaging of surfaces which reflect electrons in the electron microscope", *Z. Phys.*, vol. 83, pp. 492–497, 1933.

[SCH 06] SCHULDT T., KRAUS H.-J., WEISEL D. *et al.*, "A heterodyne interferometer for high resolution translation and tilt measurement as optical readout for the Lisa inertial sensor", *Proc. SPIE*, vol. 10567, no. 5, p. 105673M-1-7, 2006.

[SHO 49] SHOCKLEY W., "The theory of p-n junctions in semiconductors and p-n junction transistors", *Bell Syst. Tech. J.*, no. 28, pp. 435–489, 1949.

[SHO 97] SHOR P.W., "Polynomial-time algorithms for prime factorization and discrete logarithms on a quantum computer", *SIAM J. Comput.*, vol. 26, no. 5, pp. 1484–26, 1997.

[SMI 65] SMITH H.M., TURNER A.F., "Vacuum deposited thin films using a ruby laser", *Appl. Opt.*, vol. 4, pp. 147–148, 1965.

[SUT 87] SUTTON C.M., "Nonlinearity in length measurement using heterodyne laser Michelson interferometry", *J. Phys. E: Sci. Instr.*, vol. 20, pp. 1290–1292, 1987.

[SYN 28] SYNGE E.H., "Suggested method for extending microscopic resolution into the ultramicroscopic region", *Phil. Mag.*, vol. 6, pp. 356–362, 1928.

[SYN 32] SYNGE E.H., "An application of piezoelectricity to microscopy", *Phil. Mag.*, vol. 13, pp. 297–300, 1932.

[TAN 74] TANIGUCHI N., "On the basic concept of 'Nano-Technology'", *Proc. Intl. Conf. Prod. Eng. Part II*, Japan Society of Precision Engineering, Tokyo, 1974.

[TOP 03] TOPSU S., CHASSAGNE L., HADDAD D. *et al.*, "Heterodyne interferometric technique for displacement control at the nanometric scale", *Rev. Sci. Instr.*, vol. 74, pp. 4876–4880, 2003.

[VAN 04] VANDERSYPEN L.M.K., CHUANG I.L., "NMR techniques for quantum control and computation", *Rev. Mod. Phys.*, vol. 76, pp. 1037–1069, 2004.

[WEN 05] WENGER J., OURJOUMTSEV A., TUALLE-BROURI R. *et al.*, "Time-resolved homodyne characterization of individual quadrature-entangled pulses", *Eur. Phys. J. D*, vol. 32, pp. 391–396, 2005.

Index

B, C, D

E, F, G

H, I, J

L, M, N

O, P, Q

R, S, T, U

Other titles from

iSTE

in

Mechanical Engineering and Solid Mechanics

2021

DAHOO Pierre Richard, POUGNET Philippe, EL HAMI Abdelkhalak
Applications and Metrology at Nanometer Scale 1: Smart Materials, Electromagnetic Waves and Uncertainties
(Reliability of Multiphysical Systems Set – Volume 9)

2020

SALENÇON Jean
Elastoplastic Modeling

2019

BAYLE Franck
Reliability of Maintained Systems Subjected to Wear Failure Mechanisms: Theory and Applications
(Reliability of Multiphysical Systems Set – Volume 8)

BEN KAHLA Rabeb, BARKAOUI Abdelwahed, MERZOUKI Tarek
Finite Element Method and Medical Imaging Techniques in Bone Biomechanics
(Mathematical and Mechanical Engineering Set – Volume 8)

IONESCU Ioan R., QUEYREAU Sylvain, PICU Catalin R., SALMAN Oguz Umut
Mechanics and Physics of Solids at Micro- and Nano-Scales

LE VAN Anh, BOUZIDI Rabah
Lagrangian Mechanics: An Advanced Analytical Approach

MICHELITSCH Thomas, PÉREZ RIASCOS Alejandro, COLLET Bernard, NOWAKOWSKI Andrzej, NICOLLEAU Franck
Fractional Dynamics on Networks and Lattices

SALENÇON Jean
Viscoelastic Modeling for Structural Analysis

VÉNIZÉLOS Georges, EL HAMI Abdelkhalak
Movement Equations 5: Dynamics of a Set of Solids
(Non-deformable Solid Mechanics Set – Volume 5)

2018

BOREL Michel, VÉNIZÉLOS Georges
Movement Equations 4: Equilibriums and Small Movements
(Non-deformable Solid Mechanics Set – Volume 4)

FROSSARD Etienne
Granular Geomaterials Dissipative Mechanics: Theory and Applications in Civil Engineering

RADI Bouchaib, EL HAMI Abdelkhalak
Advanced Numerical Methods with Matlab® 1: Function Approximation and System Resolution
(Mathematical and Mechanical Engineering SET – Volume 6)
Advanced Numerical Methods with Matlab® 2: Resolution of Nonlinear, Differential and Partial Differential Equations
(Mathematical and Mechanical Engineering SET – Volume 7)

SALENÇON Jean
Virtual Work Approach to Mechanical Modeling

2017

BOREL Michel, VÉNIZÉLOS Georges
Movement Equations 2: Mathematical and Methodological Supplements (Non-deformable Solid Mechanics Set – Volume 2)
Movement Equations 3: Dynamics and Fundamental Principle (Non-deformable Solid Mechanics Set – Volume 3)

BOUVET Christophe
Mechanics of Aeronautical Solids, Materials and Structures
Mechanics of Aeronautical Composite Materials

BRANCHERIE Delphine, FEISSEL Pierre, BOUVIER Salima, IBRAHIMBEGOVIC Adnan
From Microstructure Investigations to Multiscale Modeling: Bridging the Gap

CHEBEL-MORELLO Brigitte, NICOD Jean-Marc, VARNIER Christophe
From Prognostics and Health Systems Management to Predictive Maintenance 2: Knowledge, Traceability and Decision (Reliability of Multiphysical Systems Set – Volume 7)

EL HAMI Abdelkhalak, RADI Bouchaib
Dynamics of Large Structures and Inverse Problems (Mathematical and Mechanical Engineering Set – Volume 5)
Fluid-Structure Interactions and Uncertainties: Ansys and Fluent Tools (Reliability of Multiphysical Systems Set – Volume 6)

KHARMANDA Ghias, EL HAMI Abdelkhalak
Biomechanics: Optimization, Uncertainties and Reliability (Reliability of Multiphysical Systems Set – Volume 5)

LEDOUX Michel, EL HAMI Abdelkhalak
Compressible Flow Propulsion and Digital Approaches in Fluid Mechanics (Mathematical and Mechanical Engineering Set – Volume 4)
Fluid Mechanics: Analytical Methods (Mathematical and Mechanical Engineering Set – Volume 3)

MORI Yvon
Mechanical Vibrations: Applications to Equipment

2016

BOREL Michel, VÉNIZÉLOS Georges
Movement Equations 1: Location, Kinematics and Kinetics
(Non-deformable Solid Mechanics Set – Volume 1)

BOYARD Nicolas
Heat Transfer in Polymer Composite Materials

CARDON Alain, ITMI Mhamed
New Autonomous Systems
(Reliability of Multiphysical Systems Set – Volume 1)

DAHOO Pierre Richard, POUGNET Philippe, EL HAMI Abdelkhalak
Nanometer-scale Defect Detection Using Polarized Light
(Reliability of Multiphysical Systems Set – Volume 2)

DE SAXCÉ Géry, VALLÉE Claude
Galilean Mechanics and Thermodynamics of Continua

DORMIEUX Luc, KONDO Djimédo
Micromechanics of Fracture and Damage
(Micromechanics Set – Volume 1)

EL HAMI Abdelkhalak, RADI Bouchaib
Stochastic Dynamics of Structures
(Mathematical and Mechanical Engineering Set – Volume 2)

GOURIVEAU Rafael, MEDJAHER Kamal, ZERHOUNI Noureddine
From Prognostics and Health Systems Management to Predictive Maintenance 1: Monitoring and Prognostics
(Reliability of Multiphysical Systems Set – Volume 4)

KHARMANDA Ghias, EL HAMI Abdelkhalak
Reliability in Biomechanics
(Reliability of Multiphysical Systems Set –Volume 3)

MOLIMARD Jérôme
Experimental Mechanics of Solids and Structures

RADI Bouchaib, EL HAMI Abdelkhalak
Material Forming Processes: Simulation, Drawing, Hydroforming and Additive Manufacturing
(Mathematical and Mechanical Engineering Set – Volume 1)

2015

KARLIČIĆ Danilo, MURMU Tony, ADHIKARI Sondipon, MCCARTHY Michael
Non-local Structural Mechanics

SAB Karam, LEBÉE Arthur
Homogenization of Heterogeneous Thin and Thick Plates

2014

ATANACKOVIC M. Teodor, PILIPOVIC Stevan, STANKOVIC Bogoljub, ZORICA Dusan
Fractional Calculus with Applications in Mechanics: Vibrations and Diffusion Processes
Fractional Calculus with Applications in Mechanics: Wave Propagation, Impact and Variational Principles

CIBLAC Thierry, MOREL Jean-Claude
Sustainable Masonry: Stability and Behavior of Structures

ILANKO Sinniah, MONTERRUBIO Luis E., MOCHIDA Yusuke
The Rayleigh−Ritz Method for Structural Analysis

LALANNE Christian
Mechanical Vibration and Shock Analysis – 5-volume series – 3rd edition
Sinusoidal Vibration – Volume 1
Mechanical Shock – Volume 2
Random Vibration – Volume 3
Fatigue Damage – Volume 4
Specification Development – Volume 5

LEMAIRE Maurice
Uncertainty and Mechanics

2013

ADHIKARI Sondipon
Structural Dynamic Analysis with Generalized Damping Models: Analysis

ADHIKARI Sondipon
Structural Dynamic Analysis with Generalized Damping Models: Identification

BAILLY Patrice
Materials and Structures under Shock and Impact

BASTIEN Jérôme, BERNARDIN Frédéric, LAMARQUE Claude-Henri
Non-smooth Deterministic or Stochastic Discrete Dynamical Systems: Applications to Models with Friction or Impact

EL HAMI Abdelkhalak, RADI Bouchaib
Uncertainty and Optimization in Structural Mechanics

KIRILLOV Oleg N., PELINOVSKY Dmitry E.
Nonlinear Physical Systems: Spectral Analysis, Stability and Bifurcations

LUONGO Angelo, ZULLI Daniele
Mathematical Models of Beams and Cables

SALENÇON Jean
Yield Design

2012

DAVIM J. Paulo
Mechanical Engineering Education

DUPEUX Michel, BRACCINI Muriel
Mechanics of Solid Interfaces

ELISHAKOFF Isaac *et al.*
Carbon Nanotubes and Nanosensors: Vibration, Buckling and Ballistic Impact

GRÉDIAC Michel, HILD François
Full-Field Measurements and Identification in Solid Mechanics

GROUS Ammar
Fracture Mechanics – 3-volume series
Analysis of Reliability and Quality Control – Volume 1
Applied Reliability – Volume 2
Applied Quality Control – Volume 3

RECHO Naman
Fracture Mechanics and Crack Growth

2011

KRYSINSKI Tomasz, MALBURET François
Mechanical Instability

SOUSTELLE Michel
An Introduction to Chemical Kinetics

2010

BREITKOPF Piotr, FILOMENO COELHO Rajan
Multidisciplinary Design Optimization in Computational Mechanics

DAVIM J. Paulo
Biotribolgy

PAULTRE Patrick
Dynamics of Structures

SOUSTELLE Michel
Handbook of Heterogenous Kinetics

2009

BERLIOZ Alain, TROMPETTE Philippe
Solid Mechanics using the Finite Element Method

LEMAIRE Maurice
Structural Reliability

2007

GIRARD Alain, ROY Nicolas
Structural Dynamics in Industry

GUINEBRETIÈRE René
X-ray Diffraction by Polycrystalline Materials

KRYSINSKI Tomasz, MALBURET François
Mechanical Vibrations

KUNDU Tribikram
Advanced Ultrasonic Methods for Material and Structure Inspection

SIH George C. *et al.*
Particle and Continuum Aspects of Mesomechanics

www.ingramcontent.com/pod-product-compliance
Lightning Source LLC
LaVergne TN
LVHW011753200926
848484LV00003B/691
9781786306876